Dictionary of Agriculture

Dictionary of Agriculture

Tahir Ali
Sandeep Kumar
Mehraj ud din Khanday

SSPH

SATISH SERIAL PUBLISHING HOUSE
403, Express Tower, Commercial Complex
Azadpur, Delhi-110033 (India)
Phone : 011-27672852, Fax : 91-11-27672046
e-mail : info@satishserial.com, hkjain1975@yahoo.com
Website : www.satishserial.com

Published by :

SATISH SERIAL PUBLISHING HOUSE
403, Express Tower, Commercial Complex, Azadpur, Delhi-110033 (INDIA)
Phone : 011-27672852 Fax : 91-11-27672046
E-mail : info@satishserial.com, hkjain1975@yahoo.com

ISBN 978-93-86200-37-2

Composed, Designed & Printed in India

Prof. (Dr.) Nazeer Ahmed
Vice-Chancellor
FISVS, FHSI, FISHRD & FCHAI

Sher-e-Kashmir University of Agricultural Sciences & Technology of Kashmir
www.skuastkashmir.ac.in

Foreword

Agriculture is the backbone of 55-60 per cent of the rural Indian population whose livelihood is directly or indirectly depends upon agriculture and allied fields but its contribution to national GDP has come down to about 14 percent thus the deficiencies on various accounts in agriculture need to be understood and necessary measures taken. The present compilation is an attempt to understand various terms commonly use in agriculture for the benefit of practitioners. Generally it is rather difficult to encompass all the knowledge of different branches to collect and compile very methodically but I hope this compilation presents a wealth of information on almost all aspects especially disciplines of agriculture. The huge number of technical terms numbering over 25,000 are exlained in simple, lucid and understandable form. I am sure the dictionary will serve as a very useful reference to all the students, scholars, teachers and extension workers. I congratulate Dr. Tahil Ali, Dr. Sandeep Kumar and Mehraj ud din Khanday for their excellent compilation and presentation.

Nazeer Ahmed

Shalimar Srinagar-190025, J&K, India
Phone: (O) 0194-2462159, 2464028 Fax: 0194-2462160 (R) 0194-2463655, 2436700 Fax : 0194-2461543
E-mail: vc@skuastkashmir.ac.in, skuastkvc@gmail.com

About the Author

Dr. Tahir Ali started his career in 1993, as soil surveyor in Water and Power Consultancy Services India Limited (WAPCOS), thereafter he worked for Maharashtra Hybrid Seed Co. Ltd., India (MAHYCO) as officer quality control from 1995 to 1998. He joined Sher-e-Kashmir University of Agricultural Sciences and Technology of Kashmir as Assistant Professor, in the Division of Soil Science on 24-11-1998 and has been working as Professor since September, 2009. He has 19 years of teaching and research experience and has guided 5 M.Sc. and 6 Ph. D students. He has published 50 research papers in various journals of national and international repute , technical bulletin , book chapters and published one book.

Dr. Sandeep Kumar completed his Ph. D in the year of 2015 in the division of soil science, Sher-e-Kashmir University of Agricultural Sciences and Technology of Kashmir and is currently working as Assistant Professor in the division of soil science Sher-e-Kashmir University of Agricultural Sciences and Technology of Kashmir and has published 10 research papers in the national and international journals.

Mehraj ud din Khanday passed B.Sc. (Ag.) from Dr. B R Ambadkar University, Agra in 2011. He did his masters in Soil Science in 2013 from Sher-e-Kashmir University of Agricultural Sciences and Technology of Kashmir. Currently pursuing his Ph. D in Sher-e-Kashmir University of Agricultural Sciences and Technology of Kashmir. In addition he has published 20 research papers in national and international Journals and one book to his credit.

Preface

Agriculture the art and science of crop production including crop protection and post harvest technology is a vast subject besides agronomy its main constituent agriculture also includes botany, horticulture, entomology, plant pathology, and even meteorology, statistics. To compile a dictionary including all these discipline will mean several thousand entries covering over a thousand pages which may not be within the easy reach of every student. Therefore this comprehensive compendium has been complied to cover most of the terminology that an agriculture graduate is likely to come across during his college career. Attempt has been made to cover all the above aspects but only important and more commonly used terms have been described to keep the size and price within limits. It is hoped that this dictionary will be of immense help not only students but also to the teachers, scientists, agriculture officers and progressive farmers of our country.

Dr. Tahir Ali
Dr. Sandeep Kumar
Mehraj ud din Khanday

Contents

1 Agronomy

A horizon: The surface horizon of a mineral soil having maximum biological activity or eluviation (removal of materials dissolved or suspended in water) or both.

A-value: Radio-chemical analysis of plants grown on soils which have been treated with fertilizes containing elements such as radioactive phosphorous may be used to calculate the phosphorous supply of the original soil.

Abiotic Factors: Physical, chemical and other non-living components of the environment such as light, temperature, moisture, nutrients and other edaphic factors.

Absolute Atmosphere: An absolute unit of pressure, equal to one million times the pressure produced on a square centimeter by the force on one dyne.

Absolute Drought: A period of at least fifteen consecutive days none of which has received as much as 0.25 millimeters of rain.

Absolute Growth Rate (AGR): The rate of increase in size of a growing plant (or part of it) in a given time under specific condition.

Absolute Water Requirement: Also called consumptive use of water. This is the quantity of water in acre-inches per crop season absorbed by the crop together with the evaporation from the crop producing land. It includes the water used by evapotranspiration including retained in the plant body.

Absolute Weed: Weeds of no use to farmers under any circumstances. These are plants which are undesirable, regardless of time and place.

Absorbed Water: Water held mechanically in a soil mass and having physical properties not substantially different from ordinary water at the same temperature and pressure.

Acid Equivalent (ae): 1. The theoretical yield of parent acid from the active ingredient content of a formulation. 2. The total organic acid content expressed in terms of the active acid 3. Herbicidal activity is accurate calculated on an acid equivalent basis. It refers to that part of a formulation that can be theoretically converted to the acid. The acid equivalent in such cases is given as active ingredient. 4. The amount of active ingredient expressed in terms of parent acid. It is used instead of or in addition to the active ingredient for certain herbicides.

Adjusted Crop Yield Index (ACYI): It is a crop yield index that takes account of the cropping intensity as shown below:

ACYI=Crop yield index x (Cropping intensity on the farm) / (Cropping intensity in the area).

Adjusted SAR: A concept that incorporates the effect of SAR and RSC through a single value. It is represented as : adj. SAR = Na 1+(8.4=pHc) (Ca+Mg)/2 or adj.SAR=SAR 1+(8.4-pHc). This equation takes care of changes in soil water composition that are expected to result due to certain combination of ions in irrigation water which will either result in deposition of lime from the soil water or will dissolve lime from the soil.

Adjuvant (Additive): 1. Chemical capable of improving herbicidal effect in some way. 2. An ingredient which when added to a formulation aids the action of the toxicant. Includes such materials as wetting agents, spreaders, adhesive, emulsifiers, dispersing agents and correctives. 3. An inactive material which when added to a plant growth regulator enhances its efficacy..

Adsali Sugarcane: Sugarcane which takes 18 months for harvesting, usually planted in June-July.

Adsorbed Water: Water in a soil mass, held by physiochemical forces, having physical properties substantially different from absorbed water or chemically combined water at the same temperature and pressure.

Adsorption: 1. The process by which chemicals are held on the surface of a mineral or soil particle. 2. The retention of one substance (the adsorbate) at the surface of another (the adsorbent). 3. The Physical gathering of toxicant molecules or ions at specific colloid-water interfaces.

Aerial Irrigation: Also called overhead or sprinkler irrigation. The conveying of water under pressure through pipes to the area to be irrigated and passed out through nozzles of sprinklers.

Afforestation: Is planning of trees where none grew previously or recently.

Ag Soil: A soil with two horizons, black A horizon over gleified g horizon due to high water table.

Agriculture: The production of plants and animals for the use of human being. This includes cultivation of soil and the feeding, breeding and managing of crops and livestock. To some extent it deals with preparation of plant and animal product for human consumption and marketing of these products.

Agricultural Drought: It occurs where soil moisture and rainfall are inadequate during the growing season to support healthy crop growth to maturity and cause extreme crop stress and wilt.

Agricultural Ecosystem/Agro-Ecosystem: It is composed of the total complex of the crops/animals in an area together with overall environment and as modified by management practices.

Agrisilviculture: A combination of forestry and agriculture crop production either simultaneously or in rotation.

Agro-ecology: The study of the relation of agricultural crops and environment.

Agroclimatic Regions: The grouping of different physical areas within the country into broadly homogenous zones based on climatic and edaphic factors.

Agroclimatology: The study of those aspects of climate which are relevant to the problem of agriculture.

Agroforestry: A collective term for systems of land use in which woody plants (trees and shrubs) are deliberately combined on the same land management unit with herbaceous crops and/or animals; either in some form of spatial arrangement or in sequence. For a land use system to fall within the concept of agro-forestry, there should be both ecological and economic interactions between woody plants and other components of the system. It may involve the integration of trees into farming systems of crops and livestock into forests. In practice, a high proportion of agro-forestry systems involve the growing of trees on what is primarily agricultural land.

Agrological Map: A map giving the agricultural properties of soils.

Agrology: The study of the properties of the soil considered from the standpoint of their relations to plants. It differs from pedology in that the letter studies the soil for itself.

Agrometerology: A science dealing with climatologic conditions which have direct relation or relevance to agriculture.

Air Filled Porosity: The proportion of a soil's volume that is filled with air at a given time; a low air-filled porosity indicates poor airation and restricted drainage.

Akiochi Soils: These are flooded soils where hydrogen sulphide is formed due to sulphate reduction and anaerobic decomposition of organic matter.

Algicide: Chemical used to destroy algal growth.

Alien Weeds. Are weeds introduced form some other part of the world. They are highly troublesome e.g., waterhyacinth, Mexican poppy, carrotgrass. Such weeds are also called Anthrophytes. Compare Native Weeds. Quarantine people need to identify alien weeds to check their movement from one country to another.

Alkali soil: A soil with a high degree of alkalinity (pH 8.5 or higher) and with high exchangeable sodium content (15% or more) with electrical conductivity less than 4 deci-siemens/m at 25°C.

Alkaline Soil: A soil with a pH above 7.0 usually above 7.3.

Alkalinity: The term used to represent the content of carbonates, bicarbonates, hydroxides and occasionally borates, silicates and phosphates in water. It is expressed in parts per million of calcium carbonate.

Alley Cropping: A farming system in which arable crops are grown in alleys formed by trees or shrubs, established mainly to hasten soil fertility restoration and enhance soil productivity.

Alluvial Soil: A soil developed from recently deposited alluvium and exhibiting essentially no horizon development or modification of the recently deposited materials.

Alternate Grazing: Grazing of two or more pastures in succession.

Ameliorants: Substances added to the soils for improving their physical and chemical properties and to help in increasing crop yields.

Amendment: Any substance which improves the physical and chemical properties of the soil form the point of view of farming, e.g. gypsum, pyrites etc.

Antagonism: 1. Opposing action of two or more chemicals such that the combined effect is less than the sum of their independent effects. 2. The loss of activity of a chemical when exposed to another chemical. 3. Negative action of chemicals used in combination so that the total action is inferior to their independent effects. Opp. SYNERGISM.

Antagonistic Effect: An interaction term which show that the total effect of a combination is smaller than the effect of the most active component applied alone.

Antitranspirants: Substances that retard or inhibit transpiration. These are of three types 1. apolar substances e.g., cetyl alcohol and oxyethylene decasonal, 2. plasitics e.g., polyethylenes, and 3. substances that lead to stomatal closure e.g., phenyl mericuric acetate. A new group under the third category are alkenyl succinic acids.

Apoplast: Nonliving cells walls forming a continuous translocation system in the plant body.

Aquaculture: Farming of the ocean, wherby such organisms as fish, algae, and shellfish are grown under controlled conditions. At present this technique is only use in near shore areas.

Aquatic: A plant that grows in water. There are three kinds: **submergent,** grows beneath the surface such as waterweed and green algae; **emergent,** grows above the water, such as cattails and water lilies; **floaters,** such as water hyacinth and water lettuce.

Aquatic Agriculture: The management of natural and/or artificial means to increase the production of fish, oysters, or other animals or plant in fresh or salt waters.

Aquifers: Formations or strata within the saturated zone below the ground surface from which ground water can be obtained for beneficial use. Also called groundwater reservoirs or water bearing formations.

Arable Crops: Annual crops which required the land to be ploughed and cultivated (unless direct drilled) and seeds to be sown, as distinct from those which yield a product without such annual activities (i.e. fruit, grass etc.).

Arable Land: Land capable of being ploughed, that is, fit for tillage as opposed to that only suited for pasture.

Arid: Applied to a climate or a region where the rainfall is barely sufficient to support vegetation; deficient in rainfall.

Aridity Index (AI): is a numerical indicator of the degree of dryness of the climate at a given location.

Aridity: 1. Area having less than 250 mm of rains per annum 2. area where the total evaporation potential exceeds the actual precipitation. 3. Inadequate rainfall to support agriculture without irrigation. The aridity index is usually expressed as : $I=P/(T+10)$ Where I is the index of aridity, P is the mean annual rainfall (mm) and T is the mean annual temperature (°C).

Arviculture: Crop science.

Asexual Reproduction: Any reproductive process that does not involve the union of gametes or sex cells; in lower forms, reproduction by spores not associated with the sexual process.

Assimilation: The process whereby photosynthates are utilized in the building up of protoplasm and cell wall.

Atmospheric Moisture: Water as it occurs in various forms in the atmosphere.

Autoallelopathy: A process of toxicity originating from affected plant itself *Also called* Autotoxicity.

Autotrophic: Capable of self nourishment by using CO_2 and carbonates as the sole source of carbon and a simple inorganic nitrogen compound for metabolic synthesis.

Auxanometer: An apparatus to measure growth in length of a plant. The amount of growth is indicated in a magnified manner.

Auxin: Any organic substance (as indoleacetic acid) characterized by its utility in low concentrations to promote growth of plant shoots along the longitudinal axis (as in the Avena test) and to promote various other effects such as root formation and bud inhibition.

Available Nutrient: That portion of any element or compound in the soil that can be readily absorbed and assimilated by growing plants. (Available should not be confused with exchangeable).

Available Soil Moisture: The amount of water in the soil at any time in excess of the wilting coefficient, expressed either as percentage by weight of dry soil, or as equivalent of water per unit depth of soil. Also called available moisture, available water.

Available Water Capacity (AWC): The quantity of water which a soil is capable of holding between field capacity and permanent wilting point. It is usually expressed on a volume basis in units of mm of water per 100 mm depth of soil or inches of water per foot depth. AWC = [(FC-PWP) x Depth)]/100. Where FC = Field capacity, PWP = Permanent wilting point and ASG = apparent specific gravity, also called available field capacity.

Available Water Holding Capacity: The capacity of a soil to hold water available for plant use. It can be calculated by the formula between field capacity and wilting point; v = volume weight of the soil expressed in grams per cubic centimeter; and d = depth of soil in centimeters. AWHC is given in centimeters.

Available Water: The portion of water in a soil that can be readily absorbed by plant roots; considered by most workers to be that held in the soil between field capacity and wilting point.

B horizon: The subsoil layer in which certain leached substances are deposited.

Barren Lands: Flat open land with vegetation limited to small trees and sparse shrubs.

Basal Dose: Amount of manures or fertilizers applied to the soil just before the crop is sown or planted or transplanted.

Basal treatment: 1. Herbicide treatment applied to the stem of woody plants just above the ground line. 2. A method of treating brush and trees with herbicides involving treatment of basal 30 cm of stem, preferably after peeling off their bark, to the point of liberal run-off.

Basic crops: Crops which because of acreage, value, climate, etc., are considered important in agricultural economy, paddy, wheat, cotton, sugarcane, etc.

Basin Recharge: That portion of the storm rainfall that is required to satisfy the demands of interception, depression storage, and soil moisture. It is the difference between rainfall and runoff and is often referred to as loss.

Basipetal Movement: The movement of chemicals with plant from the apex towards the base.

Basophile Weeds: Weed species that are characteristically alkali plants e.g. Alkaligrass, quackgrass, etc.

Benchmark Survey: A systematic study aimed at collecting data, e.g. on existing crops, varieties, yield, socio-economic constraints, before a project begins, Data collected depict the existing picture of the survey areas with regards to selected parameters and can be used to evaluate the results of the project.

Benefit/Cost Ratio: The present value of benefits divided by the present value of costs.

Bio-technology: The application of biological organisms, systems or processes to manufacturing and service industries.

Bioagent Hardiness: A good bioagent withstands elements of its new environment in which it is transplanted. This includes freedom from its parasites and predators. The bioagent should be capable to survive short or long periods of food shortage when the target weed population is brought to low level.

Bioagent: A living organism employed to control a pest.

Bioassay: 1. A test method using living organisms to determine the presence of a chemical. 2. Measurement of the potency of any stimulus by means of reaction which it produces in living matter. 3. Determination of the concentration of pesticide or growth regulator (or metabolite) by using a sensitive indicator plant or other organism.

Biological yield: Refer to total dry matter produced by a plant or a crop in a unit land area.

Biomass: Any organic matter which is available on a renewable basis, including agricultural crops and agricultural wastes and residues, wood and wood wastes and residues, animal wastes, mineral wastes and aquatic plants, usually expressed as dry weight per unit area.

Black Cotton Soil: Soil of black colour (throughout the profile) with low organic matter and a high clay content; exhibits a strong phenomenon of swelling and shrinkage, during dry periods cracks develop that may be several cm wide and more than 50 cm deep; Depending upon their depth cotton soils are classified as shallow, medium and deep black soils.

Broadcast Application: 1. An application over an entire area or field rather than only on rows, beds, middles or individual plants. 2. Herbicide application over the entire soil surface or weed infested area.

Blind Cultivation: Cultivating with a harrow, weeder, rotary hoe, or other implement to kill weeds before a seeded or planted crop has come up.

Boot Stage: The growth stage of grasses at the time the head is enclosed by the shealth of the uppermost leaf.

Border Strip Cropping: A kind of strip cropping involving growing of strips of permanent vegetation placed around the border of a field, a farm, or a body of water. The border strip may be used to control either wind or water erosion. It is especially useful to filter the silt out of runoff water before it reached between feedlots and streams.

Bound Water: 1. Water contained in a substance, in the lowest concentration that is in equilibrium with saturated air, present in hygroscopic materials. 2. Water strongly held on the surface of colloidal particles.

Broadcasting: The process of scattering of agricultural inputs, such as seed, fertilizer and manure on the surface of the soil by hand or by implements to provide uniform distribution of material over the entire field.

Broad-leaved weeds: Used in weed terminology to designate a broad group of non-grassy like plants, usually dicots.

Brush Weeds: Refers to woody plants, underwood thickets and small trees and shrubs.

Buffer Strip Cropping: A kind of strip cropping which includes strips of permanent vegetation placed between some of the cropped strips in a field. The buffer strips are often irregular in width and perhaps discontinuous so the cropped strips may remain uniform in width across variable slopes. The buffer strips usually include any unproductive soils, steep slopes or other problem spots that may be present in the field.

Buffer solution: A solution which resists change in pH by the addition or loss of hydrogen or hydroxyl ions.

C horizon: A mineral horizon generally beneath the solution that is relatively unaffected by biological activity and pedogenesis and is lacking in properties diagnostic of an 'A' or 'B' horizon.

CAM Plants: The CAM (Crassulacean Acid Metabolism) system is prevalent in desert plants where the CO_2 is fixed at night. In these species, there is diurnal fluctuation of acidity in thickened leaves. The CAM plants are adopted to environments of more or less constant aridity, e. g. *Bryophyllum calycinum* and cactus plants.

Canopy: The branches, leaves etc. formed by woody plants, some distance above the ground. Near about a continuous stratum of foliage.

Capillary Moisture: Water held by surface tension in the capillary spaces and as a continuous film around the particles, free to move under the influence of capillary forces, and available to plants. It is held with a tension 60 cm of water. It moves from thick to thin films and constitutes soil solution.

Capillary Movement: Movements of soil moisture in any direction caused by capillary tension.

Carrying Capacity: The maximum number of organisms/animals/plants the can be supported in a given area or habitat over a period of year (generally refers to the number of grazing livestock that can be maintained on a given grassland over a period of time).

Catch Crop: 1. A crop grows temporarily on land that will eventually carry a crop of different species, in order to maintain the productivity or prevent the erosion of the land prior to the establishment of the main crop 2. A crop employed to stimulate germination and growth of some specific, parasitic weed and then destroyed before planting the main crop.

Cereals (Crops): Cereals are the crop plants belonging to the grass family Gramineae which are grown for their edible starchy seed botanical known as caryopsis. The word 'ceres' means the godess of grain, e.g. wheat, rice, maize.

Check Dam: A small, low dam constructed in a gully or other water course to decrease the velocity of stream flow and thereby to minimizing channel scour and promote deposition of eroded material.

Check Irrigation: A method or irrigation where the water is applied to a field or orchard which has been divided into a series of level checks by earth ridges, the water flowing from one check to another along the slope. The checks may be 1/4 hectare to 1 hectare in size.

Cheeling: A mechanical method of weed control which involves the use of hoe with long handle. It cuts and scrapes the aboveground weed growth at the ground level and rakes it up. It is useful on annuals and biennials as they do not re grow from the plant parts left in the soil.

Chelate: An organic compound capable of holding the plant nutrient in a form which prevents it from getting tied with other elements in the soil, thus keeping it more or less in available form for the plant. The term refers to the claws of a crab illustrative of the way in which the atom is held e.g. EDTA.

Chemical Degradation: A pathway of herbicide degradation in soil which may involve hydrolys's oxidation, reduction, dealkylation, etc. The rate of degradation is proportional to its concentration. It usually beings as soon as a herbicide is applied to soil.

Chemical Fallow: The substitution of herbicide or tillage tools to control weeds on fallow land.

Chilling Injury: An Injury caused by relatively low temperatures above the freezing point.

Chlorosis: 1. With regard to herbicide, it means a slow acting effect of the herbicide on the plant over a length of time. 2. How poisonous a pesticide is to an animal (or man) after small repeated doses over a period of time.

Clay (Soil): Soil material that contains 40 per cent or more clay less than 45 per cent sand, and less than 40 per cent silt.

Cleistogamous, Cleistogamy: Descriptive of a flower that does not open and is self-pollinated.

Clone: A plant propagated vegetatively, for example by taking cutting. It has identical characteristics to its parent.

Coefficient of Variation: A measure of dispersion or relative variability equal to 100 times the standard deviation divided by the arithmetic mean; the standard deviation expressed as percentage of mean. Also called coefficient of variability. It is given by CV(%) = (Standard deviation x 100)/mean.

Cold Hardiness: A process by which a plant acquires hardiness to withstand cold conditions due to continuous exposure to low temperatures.

Collective Farming: The members of collective surrender their land, livestock and stock to the society. The members work together under a management committee elected.

Combine-harvester-thresher: A machine designed for harvesting, threshing separating, cleaning and collecting grain while moving through the standing crop in one operation.

Command Area: The irrigable land area by the canal or from a dam or an irrigation project.

Compaction: Increasing soil bulk density and decreasing porosity due to the application of mechanical forces to the soil. firming is a process of achieving of desirable degree of compaction.

Companion Crop: Any subsidiary crop grown in association with a main crop.

Conveyance Efficiency: Ratio between water received at the inlet to a block of field and that released at the projects head works; fraction usually expressed in percentage.

Conveyance Losses: Losses of water in transit from the source of supply to the point of service whether in natural channels or in artificial ones, such as cannals, distributaries, ditches, or water courses. They comprise evaporation from the water surface, seepage, and incidental transpiration by vegetation growing in the water or long the banks of natural channels, canals or watercourses. Also called transmission Losses.

Cooperative Farming: Farming by voluntarily pooling the lands and resources of individual farmers. The farming operations are managed jointly with the main objective of utilizing their available resources rationally by adopting coordinated and also between methods of farming.

Correlation Coefficient (r): The degree of correlation range from -1 to +1. A correlation coefficient of zero means that the two variables are not interrelated. An 'r' value of -1 and +1 indicates complete association. A positive (plus) correlation means that high values of one variable are associated with high values of the other. A negative (minus) correlation menas that as one variable increases the other tends to decrease.

Corrugation Irrigation: A method of surface irrigation in which small streams are allowed to flow through a series of narrow, shallow furrows in permeable soil long enough to permit the horizontal seepage from adjacent furrows to meet. These narrow, shallow furrows are some-times called corrugations or rills.

Cost Benefit Analysis: Is a tool in project cultivation or project appraisal. The essence the method is to determine all costs incurred by and all benefits received from the project throughout its life. The one with highest benefit/cost ratio should be implemented first.

Cost of cultivation: This is supplementary index to indicate the amount of capital resources needed to adopt a particular cropping system and the farmer with limited resources will prefer lower cultivation cost and cropping system with low profit over the system yielding high profit with high investment in many cases.

Cotton Bale: A pressed bale in ginning and pressing factory which weight about 178 kg of lint.

Cotton Boll: The fruit of cotton plant.

Cotyledons Seed leaves: The first leaves contained within the seed. They may or may not appear above ground when the seed grows. In most, but not all weeds the cotyledons are apparent in the seedling stage. They wither as the plant gets older.

Cover Crop: A close-growing crop grown primarily for the purpose of protecting and improving soil between periods of regular crop production or between trees and vines in orchards and vineyards. Cover crops are known to reduce the danger of soil erosion and to protect the land from leaching losses.

Creep Erosion: Movement of soil down steep slopes due to gravity.

Critical Period: 1. The shortest time span in the ontogeny of plant growth when treatment will result in maximum effect. 2. Period when the crop yield is affected due to inadequate moisture.

Crook Stage: The stage after bean seedling has broken through the soil but before the stem has become erect.

Crop Associated Weed: 1. Non-parasitic weed associated with specific crop. 2. Are crop specific weeds which are associated with the crop for several reasons. (i) need for specific microclimate, (ii) mimicry because of their similarity in morphology with host crops, (iii) ready contamination of crop seeds with weed seeds as their seeds mature as the same height and time.

Crop Bound Weed: 1. See parasitic weed. 2. Those species of weeds which usually parasitise the host crops. They depend for their survival upon their host plants for nutrition, partially or fully, e.g. *cuscuta* spp. *Orobanche spp.* and *Striga spp.*

Crop Cafeteria: Crop cafeteria is the demonstration of identified efficient crops/varieties in agrometeorological region/zone offering and opportunity to the farmer to choose a suitable crop combination commensurating with available resource and requirements. This approach is also applicable in case of production technology where a number of them are demonstrated and the farmer has a choice to choose a suitable one.

Crop Calender: A list of the standard crops of a region in the form of a calendar giving dates of sowing and various operations including harvesting to be done during the crop seasons in years of normal weather.

Crop Canopy: It is the structure of aerial vegetative parts with special reference to the size, orientation, depth, density and arrangement of leaves influencing the penetration and interception of radiant energy.

Crop Coefficient (kc): Ratio between crop evapo-transpiration (ET crop) and the reference crop evapo-transpiration (ETo) when crop is grown in large fields under optimum growing conditions. It is also referred to as the ratio of actual crop ET to pan evaporation.

Crop Competition: Two plants are in competition with each other when the growth of either one of both of them is reduced or their from modified as compared with their growth or form in isolation.

Crop Duration: Days taken by a crop from germination to maturity.

Crop Ecology: The branch of 'plant ecology' which deals specifically with the study of the inter-relation amongst crop plants and environment including management practices.

Crop Ecosystems: Cropping systems ranging between monoculture and multispecies culture of fields and garden crops, singly or in combination and their relation to environmental conditions and management practices.

Crop Equivalent Yield: It is the coversion of crop yields into one form to compare the crops grown mixed/inter-cropped/sequentially cropped. Coversion is done into monetary value/protein/ carbohydrate/intrinsic energy value.

Crop Growth Rate (CGR): It represents dry weight gained by a unit area of crop in a unit time expressed as g day^{-1} m^{-2} and is given by CGR = $w_2w_1 / (t_2-t_1)$. Where, w_1 and w_2 are plant dry weight at time t_1 and t_2 respectively.

Crop Logging: It is a graphic record of the progress of the crop. It contains a series of physical and chemical measurements indicating the general condition of the plants to suggest changes in farm management for maximizing crop yields. For crop logging sugarcane plants are sampled every 35 days and analyzed.

Crop Physiology: The branch of plants physiology which provides an understanding of the dynamics of yield development in crops in terms of growth analysis, plant development and yield characters in crop stands instead of in isolated plants.

Crop Production: Crop production is concerned with the exploitation of plant morphological (or structural) and plant physiological (or functional) responses within a soil and atmospheric environment to produce a high yield per unit area of land.

Crop Productivity: Economic yield or production of plant product of economic importance, expressed in standard units per unit area.

Crop Season: The most favourable weather condition to get better yield.

Crop System: Comprises all components required for the production of a particular crop and the inter-relationship between them and environment.

Crop Tolerance: The degree or the ability of the crop to be treated with a chemical but not injured.

Crop Water Use Efficiency (Ecu): The ratio of crop yield (Y) to the amount of water depeleted by the crop in the process of evapotranspirations (ET). It is expressed as follows: Ecu = Y/ET

Crop Weather Calendar: A pictorial form which provides information on the relationship between crops and weather conditions they face at successive stage of plant growth.

Crop Year: 1. The interval from the planting to the harvest of a crop. 2. The time of harvest of 1 year to the time of harvest of the following year.

Crop Yield Index: Refers to the overall productivity level attained on the farm in relation to the productivity level obtained in the area (locality). If the crop yield index of the farm under consideration is more than 100, it is said to be relatively more efficient than the average farm.

Crop Yield: The amount of harvest per acre, or other land measure for a particular crop. Usually per year.

Crop: Plants, carefully selected and developed over many years sown on cultivated land to produce food for man and animals or raw material e.g. barely, potatoes.

Cropping Index: The number of crop grown per annum on a given area of land times 100.

Cropping Intensity: 1. It is the ratio of the total cropped area to the actual net cultivated area, usually expressed in percentage. 2. Number of crops taken in sequence in a crop year expressed as per cent. One crop, two crops and three crop indicate 100 per cent, 200 percent and 300 percent cropping intensity respectively. Also call CROPPING INDEX.

Cropping Pattern: Yearly sequence and spatial arrangement of crop or crops and fallow on a given area.

Cropping Scheme: Cropping patterns used on a farm and their interactions with farm resources, other enterprises and available technology.

Crown: The point where stem and root joint in a seed plant.

Crude Fibre: This term includes all the insoluble forms of carbohydrates found in feeding stuffs. It is made up of cellulose, lignin and some pentosans and represents the tough fibrous parts of plants.

Crude Protein: All nitrogenous substances contained in feedstuffs which include the true proteins (composed of a number of aminoacids) and non-protein nitrogen (NPN) compounds such as amides.

Cultivar: It is equivalent to 'variety' and is defined as an assemblage of cultivated plants which is clearly distinguishable by any character (morphological, physiological, cytological, chemical or others) and when reproduced (sexually or asexually) retains its distinguishing characters. The term is derived from cultivated variety.

Curing: Natural or artificial aging of the agricultural produce brought about usually by some (dry or wet) heat treatment.

Cusec: The quantity of water flowing at the rate of one cubic foot per second. One cubic foot of water weights 62.4 lbs or 6.24 gallons.

Cyclone: A system of winds blowing around the centre of low barometric pressure. The flow round a cyclone is anticlock wise in Northern hemisphere and vice versa in Southern hemisphere.

Cytokinins: A term used for substances which promote cell division and exert other growth regulatory functions e.g., zeatin (naturally occurring) and 6-benzyl amino-purine (synthetic).

Dam: 1. A barrier usually on an important scale, across a watercourse, for the purpose of impounding water or creating a reservoir (i) to raise water level, (ii) to divert water therefrom into a conduit or channel, (iii) to create a hydraulic head which can be used to generate power, (iv) to improve river navigability by means of regulated releases of stored water, and (v) to retain debris. 2. To provide a dam; to obstruct the flow.

Datum: A fixed point or level which is taken as the basis for reasoning, deduction or measurement. For all practical purposes datum level is the zero from which land altitudes and sea depths are measured.

Day Cusec: The quantity of water in cusecs flowing for twenty four hours.

Day Length: Day length is the interval between sunrise and sunset.

Dead Furrow: An open trench left in between adjacent strips of land after completion of ploughing.

Deciduous: Plants or trees that shed their leaves at a particular season, usually winter.

Deep Percolation: 1. The moisture which penetrates below the depths from which it may not be used by plants; it represents that part of the water absorbed which exceeds the field capacity of the soil within the depths of root development. 2. In a geologic Science, the amount of water that is lost from a basin by monetary income (comparable equivalents) annually.

Diversity Index: It measures the multiplicity of farm products which are planted by computing the reciprocal of sum of squares of the share of gross revenue received from each individual farm enterprise in a single year. Thus, it is in effect an average of the number of crops harvested weighted in a particular way by the value of production of each of the crops.

Dormancy: 1. State of suspended growth of seed or other plant organ due to internal causes. 2. The characteristic of the seeds of plants, though viable, to be incapable of germinating immediately after they become mature even when placed under favorable conditions. 3. State of inhibited germination of seed or growth of plants organs. A state of suspended development. 4. A period of inactivity in spores, seeds, bulbs, buds and other plants organs in which active growth has temporarily ceased. 5. A state of seeds and buds in which they are alive but not germinated. In other words, dormancy is development suspended. Dormancy allows storage of millions of weed seeds in soil and enables them to grow in flushes over years.

Dormant: 1. Describes a plant that is resting between periods of growth. The dormant period of most plants is usually during the autumn and winter. 2. State of inhibited growth of seed or other living plant organs due to internal causes. 3. Stage of inhibited growth of seeds or other plant organs due to internal causes; temporary suspension of visible growth.

Double Cropping: Growing a second crop in one year growing season, after the first crop has been harvested from the same piece of land (overlap between crop can occur).

Drain: Outlet channels to remove excess water by surface flow or by percolation through the soil.

Drainage Requirements: The amount of water that drain must convey in a given time for satisfactory crop growth.

Drainage: 1. The removal of excess surface and groundwater from any area. 2. The mode or manner in which the surface waters find their way into streams and rivers. 3. The removal of surplus ground or surface water by means of subsurface or surface conduits, artificial or natural; also a system of such conduits. 4. The natural lines of depressions in an area through which storm water escapes to rivers and streams. 5. The removal of surplus ground or surface water by artificial means. 6. The manner in which the waters of an area are remove. 7. The area from which waters are drained; a drainage basin.

Dressing: A material such as organic matter, fertilizer, sand or lime that is incorporated into the soil. A top dressing is applied to the surface only without being dug in.

Drizzle: Liquid precipitation in the form of water drops of very small size; by convention, with diameter between 0.2 to 0.5 mm.

Drought (Agricultural): A shortage of moisture for crop growth or animal production.

Drought (Hydrological): A depression of surface and underground water levels or diminution of stream flow.

Drought (Meteorological): A rainless situation for an extended period during which some precipitation should normally have been received depending upon location and season.

Drought Avoidance: Permits a crop to grow for a longer time in a given moisture stress conditions (may be due to deep root system/uses less water).

Drought Resistance: A characteristic of plants suitable for cultivation in dry conditions no matter what mechanism they possess which provides resistance. One of the more important properties is their capacity to endure, without injury, an intense loss of water.

Drought: The deficit which results when the amount of water available in the soil is not sufficient to meet the demands of potential evapotranspiration. It is thus a condition where water need is in excess of available moisture. Also called DROUTH.

Dry Farming: A type of agricultural operation involving deep cultivation of the soil to form a sufficient reservoir for the moisture as it falls, surface cultivation to prevent or reduce evaporation; selection of drought resistant crops, it is practiced in those sections of the country where precipitation is deficient and where irrigation is, in general, not feasible because of cost of development or scarcity of water supply.

Dry Land Crops: Wheat, barley, oat, bajra, millet, raji, red gram, pigeon pea, arhar, tur, black gram, moth bean, cowpea, cluster bean, guar, sesamum, lined, flax, rapeseed, castor, safflower, mustard, toria, taramira, cotton, jute, san, alfalfa, clover, marvel grass, etc.

Dry Land Farming: 1. The practice of crop production in low rainfall area by conserving natural precipitation. Supplemental irrigation may be given from harvested runoff water. 2. The practice of crop production without irrigation where rainfall is usually deficient.

Dry Land: 1. Used in India to mean the opposite of wet lands, the latter lands are usually used for rice and are flooded. 2. Used to describe land tracts which are primarily rainfed as opposed to irrigated.

Dry Spell: A period of at least 15 consecutive days, during which not a single wetting rain has fallen.

Dust Mulch: An induced loose, fine, granular or powdery condition in the surface soil; formerly believed to be effective in reducing evaporation losses of water from soil.

Early Post emergence: After emergence of weeds but before crop emergence.

Earthing Up: The process of putting the earth or soil just near the base of stems of certain crop like sugarcane, cassava, papaya, potato, banana etc. to give support to the plants.

Eco-Farming: Potential for introducing mutually reinforcing ecological approaches to food production.

Ecological Niche: The position or functional status of an organism within its community and ecosystem, resulting in organisms' structural adaption, physiological response and specific behaviors.

Ecology: Study of the interrelationship between living organisms with each other and with their environment.

Economic Poison (Pesticides): As defined under the Federal Insecticide, Fungicide, and Rodenticide Act, economic poison 'mean any substance or mixture of substances intended for preventing, destroying, repelling or mitigating insects, rodents memetodes, fungi or weeds, or any other forms of life declared to be pest and any substance or mixtures of substances intended for use as a plant regulator, defoliant or desicant'. As so defined, economic poisons are now known generally as pesticides.

Economic Yield: Economically useful part of total dry matter produced by a plant or crop in unit land area.

Ecosystem: An ecosystem has been defined as an open system comprising plants animals, organic residues, atmospheric gases, water and minerals that are involved together in the flow of energy and circulation of matter. An ecosystem results from the integration of all living and non-living factors of the environment for a defined segment of space and time.

Edaphic Factor: Any soil factor which affects the organisms growing on or in it. Edaphic factors may be physical, chemical or biological and include pH, mineral and humus content etc.

Effective Irrigation: Is controlled and uniform application of water to crop land in required amount at required time, with minimum cost to produce optimum yields without waste of water and adverse effect on soil in form of soil salinity and water-logging problems.

Effective Rainfall: The proportion of total rainfall which is useful or utilizable for productive purposes. According to FAO it is that part of the total annual or seasonal rainfall which is useful directly and /or indirectly for crop production at the site where it falls, but without pumping. It, therefore, includes water intercepted by living or dry vegetation that is lost by evaporation from the soil surface, the precipitation lost by evapotranspiration and fraction which contributes to leaching and facilitates other cultural operations, either before or after sowing, without any harm to yield and quality of the principal crop.

Effective Root Zone: Is the depth of soil from which the plants draw water and nutrients. It should not be confused with the true root zone.

Efficiency of Irrigation: The fraction of the water diverted from a river or other source that is consumed by the crop, expressed as percent. Often applied to whole irrigation systems and takes account of conveyance losses.

Efficient Cropping Zone: It is a cropping zone in which relative yield index is 200% and relative spread indices vary from 90-200".

Electromagnetic Control of Weeds: Application of UHF energies to irrigated seed-beds have been found to destroy many weed propagules. Laser rays have also been successfully tested against water hyacinth. Electro-magnetic energy has the advantage of zero phototoxic residues in soils.

Emasculation: In bisexual flowers, the removal of stamens before they burst and have shed their pollens. Its purpose is to check self pollination and is done before crossing.

Emergence: 1. The time when the first leaves of the crop plant come through the ground. 2. Coming out of a place, such as germinating seedling breaking through the soil surface or a flower from a bud or the adult insect leaving the last nymphal skin or pupal case.

Emulsified Concentrate (Pesticide): An emulsified concentrate is an oil based liquid compound containing a high concentration of insecticide compound in a mixture in water, and wetting and sticking agent to make the material cover and adhere to the plants.

Emulsifier: 1. A surface active substance which promotes the suspension of one liquid in another. 2. A surface active agent which reduces interfacial tension and which can be used to facilitate formation of an emulsion of one liquid in another. 3. A chemical which facilitates the formation of an emulsion.

Emulsifying Agent: 1. A substance which stabilizes (reduces the tendency to separate) a suspension of droplets of one liquid in another liquid which otherwise would not mix with the first one. Also referred to as an emulsion stabilizer or emulsifier. 2. A surface active material which facilities the suspension of one liquid in another.

Emulsion Concentrate: A liquid formulation of a concentrated pesticide which is diluted with water or oil to form an emulsion.

Emulsion: 1. A dispersion of fine particles of any material in water (OW type) or less commonly in pesticide usage a dispersion of water droplets in oil (OW type). 2. A mixture in which very small droplets of one liquid are suspended in another liquid e.g. oil water when the emulsion consists of droplets of water in oil, it is known as an invertemulsion of mayonnaise emulsion. 3. A mixture of two liquids (stable or unstable held together by a third chemical or emulsifier). 4. The suspension or one liquid as minute globules in another liquid; for example, oil dispersed in water.

Energy Farming: A concept involving the farming of fast growing plants/trees for the purpose of providing biomass that can be used directly as fuel or converted into other forms of fuel or energy products.

Energy: The capacity to perform work. It exists in many different inter convertible forms such as heat, light, motion and electricity.

Enforced Dormancy: A kind of dormancy in weed seeds due to their deep placement in soil during ploughing of the field. Enforced dormancy is a non-specific character of seeds and it is caused by the absence of red light under the ground or under some dense foliage. The red light is known to induce seed germination by activating its phytochrome system comprising a responsive chromophobe blue pigment in seeds. Cultivation counters enforced dormancy by bringing the weeds to surface where they exposed to light, besides better aeration.

Enrichment Ratio of a Soil Constituent: Ratio of concentration in soil material in run-off to concentration in soil from which run-off originated.

Environment: The sum total of all the external forces or factors both biotic and abiotic that affects the physiological behavior or performance of an organism or of a group of organism.

Enzyme: A complex organic substance, a protein, produced by cells and having the power to initiate or accelerate specific chemical reactions in the metabolism of plants and animals.

Equivalent Acidity: The number of parts by weight of $CaCO_3$ that corresponds in acid neutralizing capacity to 100 parts by weight of the fertilizer.

Equivalent Weight of a Soil: The weight of a clay or organic soil colloid which has combining power equivalent to one gram atomic weight of hydrogen.

Eradication (Disease): A method of disease control in which the pathogen is eliminated after it has established.

Erodibility: The relative ease with which one soil erodes under specified conditions of slope as compared with other soils under the same conditions this applies to both sheet and gully erosion.

Erosivity: The ability of rainfall to cause soil erosion.

Evaporation Retarders: 1. Surface active substances or materials, insoluble in the substrata, forming a monomolecular film barrier on water surfaces to reduce evaporation losses from water reservoir. It refers to both the substances and the film barrier so created. 2. Methods or measures of evaporation reduction from water surfaces as windbreak, shading etc. 3. Also called EVAPORATION RETARDANTS.

Evaporation: The process by which water is changed from the liquid or the solid state into the gaseous state through the transfer of heat energy.

Evapotranspiration (ET): Water transpired by vegetation plus that evaporated from the soil. Approximate synonym: consumptive use, although consumptive use also includes water in the plant (plant water is less than 1 per cent ET).

Exchangeable Sodium Percentage (ESP): It is the degree of saturation of the soil exchange complex with sodium expressed in percentage. ESP = Exchangeable Na (m.e./100 g soil)/cation exchange capacity (m.e./100 g soil) x 100.

Exotic (plant): A newly introduced plant not native to a place.

Extensive Farming: Method of farming in which a large amount of land is used to raise stock and produce crops.

Facultative Weed: 1. A weed found growing both wild and in association with man. The prickly pear and various species of wild onion, trench etc., are examples of facultative weeds. 2. Weed of wild community origin, spacing sometimes to crop field

Fallow: It is the practice of allowing crop land to lie idle during a growing season to build up the soil moisture and fertility contents, so that a better crop can be produced the following year. It is usually worked periodically to control weeds and improve moisture infiltration.

Fallow land Weeds: Weeds that occur in land that is fallow.

Farm Crops: Economic crops grown on a field scale on a farm as different from truck crops, fruit crops etc.

Farm Irrigation Efficiency: Is the ratio expressed in percentage of irrigation water available for crop production to that delivered at the farm head gate.

Farm Management : 1. The science which considers the organization and operation of the farm from the point of view of efficiency and continuous profit. 2. Utilization of sound principles in the selection, organization and conduct of an individual farm business for the purpose of obtaining the greatest possible profit.

Farm Pond: Are the bodies of water made either by constructing a dam or an embankment across a water course or by excavating a pit or dug out.

Farming System Research: Is a highly location specific research with multi and inter disciplinary in nature and uses a whole farm approach for improved technologies to enhance and stabilize agricultural production. The research strategy includes base data analysis, on centre research and on farm research. This is the final evaluation of system in the real world situation of the farmer.

Farming: The science and art of crop production. The direct application of the multiple aspects of agronomy.

Fertigation: Application of fertilizers with irrigation water. Its applicability depends on external conditions, such as the type of soil and crop, method of irrigation used, water quality, types of fertilizer available, and general economics when compared to other ways of fertilizer application.

Fertilizer Additive: Fertilizer materials added to spray to enhance absorption of aqueous herbicides e.g. ammonium sulphate and dihydrogen phosphate. Fertilizers apparently solublize the cuticle and increase the permeability across the leaf cells.

Fertilizer Mixture: A product obtained by physically mixing different fertilizers so as to contain more than one of the three major nutrients, viz nitrogen, phosphorus and potassium.

Fertilizer-Grade: An expression showing the legal guarantee of its available plant nutrients expressed as percentage of plant nutrients in a fertilizer, e.g.a 10 - 5 - 5 grade of fertilizer indicates 10 per cent nitrogen, 5 per cent phosphoric acid (P_2O_5) and 5 per cent potash (K_2O).

Field Capacity: The moisture percentage, on a dry weight basis, of a soil after rapid drainage has taken place following an application of water, provided there is no water table within capillary reach of the root zone. This moisture percentage usually is reached within one to three days after an irrigation, the time interval depending on the physical characteristics of the soil.

Field Irrigation Efficiency: Overall efficiency in the irrigation of a given field or plot of land consists in examining together moisture distribution factors, moisture storage capacity factor and field water application efficiency. For some purposes these terms must be used individually, but to tell whether a truly good job has been done of irrigation, they must be examined together. Also called irrigation efficiency.

Field Water Use Efficiency (Eu): The ratio of crop yield (Y) to the total amount of water used in the field (*WR*). It is expressed as follows: Eu = *Y/ WR* Crop yield can be defined either in terms of growth i.e. dry matter produced or in terms of marketable product.

Fifteen Atmosphere Percentage: The moisture percentage in a soil sample which has been wetted and then brought to equilibrium on a cellulose membrane in the pressure membrane apparatus at a pressure of 221 pounds per square inch (15 atmosphere). This characteristic moisture value for soils is expressed in terms of moisture percentage, dry weight basis, and lies in the wilting range for many soils.

Foliar Fertilization: Fertilization of plants, or feeding nutrients to plants, by applying chemical fertilizers to the foliage usually in the form of spray.

Forage Crop: Crops grown primarily for livestock feed, to be either harvested for hay, silage or green feed or harvested by grazing animals.

Forecast: A statement of anticipated meteorological conditions for a specific place (or area) and period of time.

Frost Injury: Injury to the crop plants caused by temperature below the freezing point of water (0°C). Crops can be protected against frost injury by spraying water during the time when actually frost is occurring or by applying irrigation not long before a possible night forest.

Furrow slice: The mass of soil cut, lifted and thrown to one side of an implement.

Geotropism: A growth movement in response to gravity. Roots are positively geotropic, shoots negatively geotropic.

Germicide: An agent capable of killing germs usually pathogenic micro organisms.

Germination: 1. The resumption of growth of a seed spore. Usually recognized by rupture of the seed coat or spore wall and appearance of the radicle, hypocotyl or thallus. 2. The development of a normal seeding from a seed.

Gibberellins: Growth hormones, not identical with auxins, which markedly increase the elongation of stems in many plants, they also affect other processes.

Ginning (Cotton): The process of separating lint from seed cotton in ginning factory.

Ginning Percentage: The weight of cotton lint obtained from seed cotton expressed in terms of percentage of seed cotton or percentage of lint obtained from 100 unit of seed cotton by weight.

Glass House: Building constructed almost entirely of glass providing a sheltered environment, and warmer conditions than the open air for the growing of flowers and vegetables. Artificial heating may also be provided. It is also know as a green house.

Gossipol : A phenolic pigment in cotton seed that is toxic to some animals.

Grade (fertilizer): The minimum guarantee of its plant nutrient content expressed as whole numbers in terms of total nitrogen (N) available phosphoric acid (P_2O_5) and water soluble potash (K_2O).

Grain: An indehiscent simple fruit of a grass in which the seed coat is united with the pericarp.

Grassland: The land on which graminaceous species represent the dominant, if not the exclusive vegetation. Grassland is intermediate in status between forest or woodland on one hand and the desert on the other.

Grassland Farming: Farming system that emphasizes the importance of grasses and legumes in livestock and land management in which the legumes are the key stones and grasses form the backbone of the grassland farming.

Gravitational Water (L.gravis=heavy): Water which moves downward through soil as a result of the earth's gravitational force upon it.

Grazing Capacity: The number of animals a specified pasture will carry for a specified period.

Green House Effect: The effect of short wave radiation can easily pass through the atmosphere to reach the earth's surface while the outgoing radiations from the earth are not permitted to pass through the atmosphere especially when it is cloudy.

Gross Cropped Area: The total area covered with crops during the year. In case two or more crops are grown on the same land during the year, the same area is counted twice or more

Gross Irrigation Requirement (GIR): Is NIR plus losses in water application and other losses. GIR can be determined by considering the appropriate losses at various stages of the crop.*GIR*= Net irrigation requirement/Field efficiency of the system, Here, it should be noted that no method of irrigation can be completely efficient since not all the water applied enters the soil and held by the root zone. Unavoidable losses such as seepage, percolation, surface runoff may occur, necessitating a sort of second best evaluation of efficiency.

Ground Cover: Protective vegetation such as provided by a solid seeded stand of forage species or the residue from the previous crop.

Ground Water: The water that occurs in the zone of saturation, from which wells and spring or open channels are fed. This term is sometimes used to include also the suspended water and is loosely synonymous with sub-surface water, underground water or sub-terranian water.

Growing Period: The duration, in days, of the period when both temperature and soil moisture permit crop growth (growing season, growth cycle).

Growth Analysis: Method of analyzing growth through plant sampling technique. By this method, observations are recorded periodically on various plant characteristics (height, flowering time, ear bearing tillers etc.) and also sampled at varying intervals (weekly or fortnightly) when the different parts such as stem, root, leaf, fruit, seed etc. are separated, weighed and measured. From these, the physiological parameters such as growth rate are calculated and these are further analyzed in term of contributing component parameters like net assimulation rate and leaf area.

Growth Regulators: 1. These are organic compounds other than nutrients which in small amounts promote, inhibit, or otherwise modify the physiological processes in plants. 2. A pesticide chemical which increase, decrease, or change the normal growth or reproduction of plant. 3. An organic substance effective in minute amounts for controlling or modifying plant processes

Growth Substance: An organic substance effective in minute amount in modifying specifc plant processes.

Guard Cropping: Means growing main crop in the centre surrounded by hardy and thorny crop, such as safflower around pea, mesta around sugarcane and sorghum around maize.

Guttation : Exudation of xylem sap usually through structures called hydathodes present at the tips of veins of leaves. This occurs in plants growing under moist and humid conditions.

Gynophore : A stalk or stipe bearing an overy or fruit.

Gypsum Block: Made up from plaster of paris with the help of two electrodes parallel to each other in a medium of gypsum block. It is a resistance unit which is used for measurement of soil moisture in 'situ'. The resistance reading is about 400-600 ohms at field capacity and 50000-75000 at wilting point.

Gypsum Requirement: The quantity of gypsum or its equivalent required to reduce the exchangeable-sodium percentage of a given alkali soil to an acceptable level.

Habitat: A specific kind of living space or environment where an organism lives or could be found.

Hailstorm: a storm, often prolonged and severe, consisting largely of hail or frozen raindrops ranging in diameter from 5 mm to 10 mm or more, the ice particles are precipitated either separately or as aggregates of irregular size and shape.

Half Life: Time needed for half the pesticide to disappear from the scene, usually soil.

Hard Pan: A hard compact layer at the surface of the soil or at a lower horizon within the soil profile. It may consist of concentrations of particles transported by soil solution of chemical compounds precipitated from solution e.g. silica, sesquioxides or calcium carbonate; hardness not changing appreciably with changes in water content. Pieces of hard layer do not slake in water.

Hard Water: Water which contains certain minerals, usually calcium and magnesium sulphate, chlorides, or carbonates in solution to the extent that a curd or precipitate rather than a lather occurs when soap is added. Very hard water may cause objectionable precipitates in some herbicidal sprays.

Hardening: Increase in the tolerance of a plant tissue to stress or adverse conditions.

Harvest Index: The -ratio of grain yield to biological yield. It is a factor less than unity, say 0 to 0.55, but some workers prefer to use harvest index (%) in which they express the index as 0 to 55 per cent. It is expressed as: Harvest Index = Grain Yield/Biological Yield or Total Yield. The determination of harvest index is usually based on a harvest at maturity. The index permits an analytical interpretation of environmental and genotypic influences than is possible from grain yields alone. Harvest index is usually derived from measurements of biological yield and grain yield, but sometimes the biological yield may usually be derived from published values of harvest index and grain yield. The depression in harvest index due to nitrogen or a high population density in a well-watered environment is probably an effect of heavy vegetable growth on light relationship within the canopy.

Hazard: The probability that toxicity will result following a given exposure. Hazard presented by a compound frequently depend more on how it is used than on how toxic it is.

Heavy Soils: A term used for clayey or fine textured soils.

Heavy Water (D_2O): Water in which hydrogen has been replaced by deuterium (hydrogen isotope of mass 2).

Herbarium: Collection of plant specimens that have been taxonomically classified, pressed, dried and mounted on sheet of herbarium paper.

Herbicide Combinations: Involves mixing of two or more herbicides (i) to broaden the spectrum of weed control, (ii) to get synergistic action, (iii) to prevent rapid detoxification of active herbicide component, and (iv) to improve herbicide safety.

Herbicide Equivalent: The possible or theoretical yield of herbicide compound for the active ingredient in a formulation (applied especially to compounds not derived from acids).

Herbicide Glove: The gloves designed to allow safe application of specially effective herbicide to tough-to control weeds.

Herbicide Rotation: 1. System of using different kinds of herbicides in recurrent succession on same land. 2. The practice of following a systematic, rotational sequence herbicide use in the same field. It may involve either (i) change of a herbicide in the same crop, or (ii) change of crop in the same field so that a different herbicide could be used. The chief objective of herbicide rotation is to curb a wide spectrum of weeds.

Herbicide Selectivity: The phenomenon wherein a chemical kills the target plant species in a mixed plant population without harming or only slightly affecting the other plants.

Herbicide: 1. A chemical used for killing or inhibiting the growth of plants. 2. *Syn* for weed killer. 3. A chemical which kills or severely damages plants.

Heterotrophic: Capable of deriving energy for life processes only from the assimilation of organic carbon compounds and incapable of using carbon dioxide as the sole carbon source for cell synthesis. Contrast with autotrophic.

Hidden Hunger: It is a term used to describe a plant that shows no obvious symptoms yet the nutrient content is not sufficient to give profitable yield.

High Analysis Fertilizer: A fertilizer containing not less than 25 per cent of the major plant nutrients, namely, nitrogen, phosphorus (as P_2O_5) and potassium (as K_2O).

High Volume: Spray application of more than 750 liters per hectare.

High Yielding Variety (HYV): Are varieties which respond to higher levels of modem inputs like fertilizers and give higher yields.

Hill Placement: This refers to applying fertilizer either in bands or near the plants. This method is usually employed when relatively small quantities are to be applied. Nitrogenous fertilizer is applied in pinches at the base of each plant of cotton, maize, cauliflower, cabbage, brinjal, tomato and other vegetable crops by this method.

Hormone: A growth-regulating substance occurring naturally in plants or animals; also refers to synthetic chemicals with growth-regulating activity. However, these are more correctly caned synthetic regulators, and not hormones.

Horse Power (Effective): In any system of which the engine is only one element, the effective horse power is the actual horse power given by or to the system. Thus, in a tractor, the effective horsepower given by the tractor to the implement is: HP = [Draw Pull (Ibs.) x Speed ft/min] /33,000

Host Specificity: Biocontrol agents for weed control should not attack other plant species, particularly the economic ones.

Host: Organism such as plant, that harbours a parasite such as a disease causing pathogen or an insect that feeds upon it.

Hull (Rice): The rice caryopsis is surrounded by a hull (husk) composed of two modified leaves, the palea and larger lemma.

Humectants: Are substance used to improve absorption of herbicides by raising the humidity between the spray film and leaf surface.

They operate on the principle that higher humidity aids in rapid and greater absorption.

Humidity: Moisture or dampness in the air. Weed killers are often comparatively more effective under moderately humid conditions. In areas or at times when humidity is very low high herbicidal rates or high volumes of carrier may be required because sprays dry more quickly and absorption is poor.

Hybrid: A plant produced by the cross fertilization of two species or variants of a species.

Hydraulic Conductivity: The coefficient "K" in Darcy's law: v = Ki. Where v is the velocity of seepage and *i* is the hydraulic gradient. Value of *K* depends on properties of the fluid, such as viscosity as well as that of the porous medium, and reflect any interactions of the fluid with the porous medium such as swelling of soil and its attendant reduced porosity.

Hydrology: The scientific study of the water found on the earth's surface, in its subsurface, and in the atmosphere, the science dealing with the properties, distribution and circulation of water and snow.

Hygroscopic Moisture: Moisture that is held in the soil in equilibrium with atmospheric water vapour at the ground surface. Hygroscopic moisture in a soil sample can only be driven off by heat.

Hygroscopic Water: Water so tightly held by the attraction of soil particles that it cannot be removed except as a vapour by raising the temperature above the boiling point of water. It is unavailable to plants.

Ideotype: Refers to plant type in which morphological and physiological characteristics are ideally suited to achieve high production potential and yield reliability.

Immature Soil: A soil with indistinct or only slightly developed horizons because of the relatively short time it has been subjected to the various soil-forming processes. A soil that has not reached equilibrium with its environment.

Immunity: Natural or acquired resistance of a plant to active infection by the pathogenic microorganisms or to the adverse effects of a particular parasite.

Impervious: 1. A term applied to a material through which water cannot pass or through which water passes with great difficulty 2. Resistant to penetration by fluid or roots.

Implements: The equipment used on farm to carry out different farm operations either with the help of tractor or bullock power.

Improved Seed: The genetically and physically pure seed of an improved crop variety. Its different categories are nucleus seed, breeder's stock seed, foundation seed, registered seed and certified seed.

In Situ: In the natural or original position or location.

Incipient Wilting: Small loss in turgidity during warm weather, even when soil is moist. It may not lead to any visible stress symptoms like rolling or folding of leaves and is usually reversible at night.

Incompatibility: Inability of one substance to mix with other without undesirable reaction e.g. with reference to insecticides, herbicide, fungicides where incompatibility involves loss of efficiency on mixing.

Indeterminate Growth: The flowers are borne on lateral branches, the central stem continuing vegetative growth with blooming continues for a long period like in cotton.

Indicator Plant: Indicator plant is one which reflects specific growing conditions either by its presence or character of growth. Such plant indicates water stress earlier than main crop plants.

Indigenous: An organism that is native to a particular habitat, as distinct from one introduced from outside the area.

Indole Acetic Acid (IAA): A naturally occurring growth regulator, the major auxin found in plants.

Induced Dormancy: A dormancy in weed seeds resulting from some sudden physiological changes in normally non dormant seed under the impact of market rise in temperature and/or carbon dioxide content of soil, low oxygen pressure, water logging, etc.

Industrial Weed Control: Control of all vegetation on industrial and similar sites including factories, special installations, railways, etc.

Ineffective Rainfall: Is that portion of the total annual or seasonal rainfall which is lost by surface runoff, deep percolation beyond root zone, the moisture remaining in the soil after the harvest of the crops and is not useful for the next crop. The period considered here is from the start of the first tillage operation till the harvest and not only the actual crop growing period.

Infiltration Rate: The maximum rate at which a soil, in a given condition at a given time, can absorb water. Also the maximum rate at which a soil will absorb water impounded on the surface at a shallow depth when adequate precautions are taken regarding border or fringe effects. Defined as the volume of water passing into the soil per unit of area per unit of time.

Infiltrometer: A device by which the rate and amount of water percolating the soil is determined by measuring the difference between the amount of water applied and the which runs off.

Infra-red Thermometer: Used to measure leaf and canopy temperature of plants. It collects the thermal energy emitted by the plant surface by fixed focus optics and converts into electrical signal by a sensitive detector which is electrically conditioned for reading in temperature units. It is quite fast and collects energy at the speed of light.

Inoculation: The artificial introduction of microorganisms on or into a medium, living' system.

Inorganic Nitrogen (Soil): Nitrogen in inorganic form viz. exchangeable ammonium, nitrite, nitrate, fixed ammonium, nitrous oxide and elemental nitrogen.

Insolation: Amount of direct solar radiation incident per unit on a horizontal area at a given level.

Integrated Fertility Management: Technologies for management of fertilizing the crops with the objective to produce optimum crop yields with minimum fertilizer inputs taking in to consideration ecological and socio-economic resources available for supply to the crop under a given agro-ecosystem, without deteriorating the soil fertility, such as application of organic manures, green manures, blue-green algae and biofertilizers along with use of inorganic fertilizers for achieving higher yields and also improving the soil productivity.

Intensive Cropping: The practice of producing maximum crop yield from given area by special soil and crop treatment, and providing all factors in the quantities needed and at the right time considered conducive to high yields.

Intensive Farming: Method of a farming in which aim is to produce the maximum number of crops per year of high yield, from amount of land available and to maintain a high stocking rate of livestock.

Intercropping: refers to growing of two or more generally dissimilar crops simultaneously on the sample piece of land, base crop necessarily in distinct row arrangement. The recommended optimum plant population of the base crop is suitably combined with appropriate additional plant density of the associated crop, and there is crop intensification in both time and space dimensions.

Intercrops: The crops raised in an orchard or other widely spaced crops for increasing the income from the same piece of land. e.g. short duration vegetables, pulses, oilseeds etc.,

Interculture: Operations to be carried out after sowing or planting in between crop rows. Includes operations like gap filling, thinning, weeding, mulching and top dressing. May be done by manual labour or by bullock or machine-drawn implements.

Irrigated Dryland: Areas with well-drained soils where water does not accumulate and irrigation is used mainly for dryland crops such as wheat, maize, potato, and other cash crops. Dryland rice may also be grown with intermittent type irrigation.

Irrigation Cycle: The frequency of irrigation as determined by the capacity of the soil and the periodical water requirements of the crop.

Irrigation Period: The number of days that can be allowed for applying one irrigation to a given area during the peak consumptive use period of the crop being irrigated. It is the basis for irrigation system capacity and equipment design. The irrigation system must be so designed that the irrigation period is not greater than the irrigation frequency. Irrigation Period = (Net amount of moisture in soil between start of irrigation and lower limit of moisture depletion) / (Peak period moisture) - (use rate of crop).

IW/CPE Ratio: Based on the climatological approach, scheduling of irrigation water is done and the depth of irrigation water is fixed for different crops. IW/CPE = Irrigation water (mm/cm)/ Cumulative pan evaporation (mm/cm).

Jhum Cultivation: The slash-and-burn type of shifting cultivation in the hill tracts of Bangladesh and Assam.

Kernel: Edible seed portion of rice, maize, sorghum etc.

Kharif Crops: Crops grown during the main monsoon season (April to September), such as Kharif sorghum, maize, bajra, rice, cotton etc.

Kreb's Cycle: A cycle of reactions which plays a central role in the metabolism of many heterotrophic organisms, capable of respiratory mechanisms. It involves an enzyme system which converts pyruvic acid to CO_2 in the presence of 0_2 with concomitant relea.se of energy in the form of ATP molecules.

Land Equivalent Ratio (LER): It denotes the relative land area under sole crop required to produce the same yield as obtained under a mixed or an intercropping system at the same management level. It is calculated as sum total of the ratios of yield of each component crop in an intercropping or a mixed cropping system to its corresponding yield when grown as a sole crop thus: LER = Yab/Yaa + Yba/Ybb, where, Yab is the yield of species 'a' in association with species 'b' and Yba is the yield of species 'b' in association with species 'a', Yaa and Ybb represents the pure stand yield of species 'a' and 'b' respectively.

Law of Constant Return: When application of fresh doses of productive resources results in an equal returns due to each successive doses, is known as law of constant returns.

Law of Diminishing Returns: Increase in capital and labour employed in the cultivation of land, caused, in general a less than proportionate increase in amount of produce.

Law of Minimum: It states that when a character such as yield is controlled by number of separate factors, the rate of the process is limited by the factor which is present in relative minimum amount.

Lay Farming: Growing grasses in rotation with agricultural crops.

Leaf Area Duration (LAD): is a measure of the ability of the plant to produce and maintain leaf area and is obtained by integrating the leaf area index over crop growth period. It is usually expressed in days or weeks. LAD=LAI x M. where, LAI = mean leaf area index, M = number of weeks in the crop growth period/ season.

Leaf Area Index (LAI): The area of one side of leaves per unit area of land surface. LAI = Total Leaf Area / Ground Area.

Leaf Area Ratio (LAR): The ratio of leaf area to dry weight of plant. It is calculated by the following equation.

$$\frac{L_2 - L_1}{(\log_e L_2 - \log_e L_1)} \times \frac{\left[\log_e W_2 - \log_e W_1\right]}{W_2 - 1}$$

Where L_1 and L_2 are total leaf area at time t_1 and t_2 respectively. W_1 and W_2 are dry weights during the same period. It can be expressed as cm^2/gm.

Legume: 1. A type of dry fruit developed from a single carpel and when mature dehiscing along two sides (sutures). 2. Any member of the pea family (Fabaceae), e.g. beans, peanuts and alfalfa.

Ley Farming: System of growing grass in rotation with agricultural crops; extensively used in U.K. and U.S.A.; helps improve soil fertility and provides good fodder to animals; also helps build up soil structure and prevent erosion. Can be conveniently adopted in places like Nilgiris abounding in good pasture land and having suitable climate.

Light Reaction: The photolysis of water, the photochemical reduction of pyridine nucleotide and photophosphorylation aretogether called as light reaction of photosynthesis.

Light Soil: A soil that is easy to work implements in or easy to cultivate as it lacks cohesiveness, such as sandy and sandy loam soils.

Limiting Factor: A factor which is in short supply or exceeds the optimum limits for an organism. A limiting factor may operate both when deficient or in excess.

Low Land Rice: Rice grown after wet land preparation of field (based on land and water management) /rice grown with 5-50 cm standing water (based on water regime).

Low Volume Spraying: Spraying, when spray material deposited in small drops do not form pools and obtained by spraying liquid upto 200 litres per hectare.

Low-Analysis Fertilizer Materials: The fertilizers containing a low percentage of plant nutrients, usually less than 30 per cent.

LSD (Least Significant Difference) ICD (Critical Difference): A value used to compare the differences among a set of averages as an aid to detect whether the differences are real or merely due to chance.

Macro-Pores: Soil characterizing size of pores having diameter of 100 microns or more which maintain high air circulation and increase draining of soil water and nutrients.

Main Crop: The principal or most important crop grown in a farm or area.

Manometer: An instrument for measuring the pressure of fluids; a pressure gauge or vacuum gauge. The pressure of the -fluid is balanced against a column of mercury, or other standard fluid is balanced against fluids in a U-tube or against the elastic force of a spring, an elastic diaphragm, or the like.

Marginal Farmer: A farmer with a holding of size not more than one hectare of irrigated land. In case of other areas this limit is raised taking into account the availability of irrigation facilities, the soil type, the cropping pattern and other relevant factors.

Marginal Land: Land that is sufficiently fertile to yield a return that just coves the cost of production. Marginal land will be abandoned if the price obtained for its products falls or the cost of production increases.

Marsh: Periodically wet or continually flooded areas with the surface not deeply submerged. Dampness is due to impermeable nature of the soil and poor drainage. Vegetation consists of hydrophytes like sedges (*Cirrus spp.*) cattails (*Typha spp.*) and rushes etc. Sub-classes are fresh water marsh and marine marsh. The term is often used synonymously with swamp.

Matric Potential: Is the amount of work that must be done per unit quantity of pure water in order to transport reversibly and isothermically an infinitesimal quantity of water from a pool containing a solution identical in composition to the soil water at the elevation and the external gas pressure of the point under consideration to the soil water.

Maximum Cropping: The highest possible production per unit area per unit time without considering cost of production or net return.

Meadow: Area covered with grasses and/or succulent forage legumes grown primarily for hay.

Mean Sea Level: The mean plane at which the tide oscillates; the average height of the sea for all stages of the tide. At any particular place, it is derived by a verging .the hourly tide heights over a 19 year period

Mechanical Analysis: The determination of the relative distribution of the size group of ultimate soil particles. The important steps in mechanical analysis are 1. separation of all particles from each other so as to cause complete dispersion into ultimate particle 2. measuring the amounts of each size group in the sample.

Mechanical Impedance: The mechanical resistance of the soil to the movement of plant roots or tillage tools.

Mechanized Farming: It is the farming in which machine drawn implemented area used for the reduction of labour requirement or elimination of manual work, timeliness of operations and improved quality of husbandry resulting in higher output and better quality of produce for increased profit.

Method of Irrigation: Method of operating an irrigation system to convey water from the source of supply and to distribute it according to crop requirements to each field served by the system.

Mho: A Siemens, a unit of electrical conductance, the reciprocal of the resistance in ohms.

Microclimate: is the local climatic condition near the ground or area around plants (up to 2 m height), resulting from the modifications of the general climatic conditions by local difference in relief, exposure and cover etc.

Micrometeorology: The science concerned with the detailed examination on a micro scale of the physical and meteorological processes taking place within a small distance, for example, between the top of a plant to its roots.

Micron: One millionth of a meter (10^{-6} m) or one thousandth of a Millimeter (10^{-3} mm/m).

Mid-season Correction: It is a contingent management practice to overcome unexpected and or unfavorable weather conditions e.g., thinning or population adjustment according to the availability of moisture in the soil during the dry spell.

Millibar: A subunit of pressure, being one-thousand of a bar; in meteorology, it is equal to a force of 1000 dynes per sq. cm. The values of atmospheric pressure are now usually expressed in millibars, and 1013 millibars is standard atmospheric pressure.

Milliequivalent: One thousandth of an equivalent.

Millimicron: The unit of length used for measuring wavelength of radiation. It is one millionth part of a millimeter, abbreviated asum. 1 mu = 10^{-6} mm, 1 mu = 10° A Millimicron is also known as nanometre (nm).

Mineral Soil: A soil consisting predominantly of and having its properties determined predominantly by mineral matter. Usually contains 20% organic matter but may contain an organic surface layer up to 30 cm thick.

Mineral Water. Water that contains a high quantity of mineral salts, such salts usually being of uncommon occurrence.

Minimum Tillage: The soil manipulation necessary to meet the minimum tillage requirements for crop production.

Minor Watershed: Part of watershed which may be selected as a suitable area for detailed planning and application of soil and water conservation measures in cooperation with relatively small local groups of landowners and operators, or local agencies as drainage districts or municipalities.

Mixed Farming: Farming system which involves the raising of crops and rearing of animals and/or poultry. Mixed farming is based upon the principle that land should support animal and animal should support land.

Mixed Fertilizers: Mixed fertilizers consist of individual or straight fertilizer materials blended together to permit application in the field in one operation. **Mixed fertilizers supply**. Two or three major plant nutrients. The percentages of these nutrients are expressed as fertilizer grade, like 10 - 5 - 5.

Mobile: Movement of element from one place to other parts, such as essential elements, nutrient, water etc. Mobility varies with the element.

Moisture (Soil): Particle holds water on its surface in the form of a thin film with a certain force. Soil pore space contains both air and water in varying proportions. Soil moisture expressed in percentage on even dry basis either on weight basis or on volume basis. The terms soil moisture and soil water are used generally as synonyms.

Moisture Availability Index: It is the ratio of assured rainfall (weekly or monthly) at 50 per cent probability level to potential evapotranspiration of the corresponding period.

Moisture Deficit Index (MDI) : This is the, measure to estimate dryness of a region and is expressed by the formula. MDl = {(P-PE PE)/ PE} x 100 Where

MDI= Moisture deficit index P = rainfall (mm) PE=Potential evapotranspiration (mm) (MOl ranges between - 20 to 80).

Moisture Index: 1. The ratio of the plasticity index to the difference between the natural water content and the plastic limit of a soil. 2. The difference between precipitation (P) and potential evapotranspiration (PE) divided by potential evapotranspiration usually expressed in a percentage.

Moisture Percentage: 1. **Dry weight basis**: The weight of water per 100 units of weight of material dried to constant weight at a standard temperature. 2. **Depth basis**: The equivalent depth of free water per 100 units of depth of soil. Numerically this value approximates the volume of water per 100 units of volume of soil.

Molal Concentration: The number of gram molecules of the solute present in 1000 grams of solvent, usually watlar.

Molar Solution: A solution in which one gram molecular weight of the substance is dissolved in the solvent and made upto one litre.

Mole: A unit of mass numerically equal to the molecular weight of the substance.

Mollisols: Soils with nearly black organic rich surface horizons and high supply of bases. They have mollic epipedons and base saturation greater than 50% in any cambic or argillic horizon. They lack the characteristics of vertisols and must not have oxic or spodic horizons.

Monoculture : The practice of growing same crop year after year on a particular piece of land.

Monsoon: A wind blowing with great persistence and regularly at a definite season of the year.

Mulch Farming: A system of farming in which organic residues or other materials are neither ploughed into the soil nor mixed with it, but are left on the surface to serve as mulch.

Mulch Tillage: A crop and soil management practice that utilizes the residual mulches of the preceding crop by leaving a percentage of this vegetal residue on or near surface of the ground as protective mulch. Also called stubble mulching, mulch farming.

Mulch: 1. A layer of organic or inorganic material spread on the soil surface around plants to conserve moisture and suppress weeds. Organic mulches also add nutrients to the soil. 2. Any material such as straw, plant residues, leaves loose soil or plastic film placed on the soil surface to reduce evaporation, erosion or to protect plant roots from extremely low or high temperature.

Multiple Cropping Index: An indicator of the number of crops harvested per year from the same parcel of land; computed as annual harvest area divided by cultivated area and multiplied by 100. The multiple cropping index is generally higher with smaller farm size. The intensity of land use is expected to increase with decrease in farm size.

Multiple Cropping: Growing two or more crops consecutively on the same field in the same year. In essence, represents a philosophy of maximizing crop production per unit area of land within a calendar year with minimum deterioration of soil.

Mycorrhiza: The morphological association, usually symbiotic of fungi with the roots of seed plants. The association is referred to as ectotrophic the fungal hyphae occur on the root surface and penetrate only the intercellular spaces and as endotrophic when the hyphae occur mainly within the cells of the host plant.

Native Weeds: Are weeds occurring within the geographical limit of their origin. Compare ALIEN WEEDS.

Net Assimilation Rate (NAR): It is increase in dry weight of plant per unit leaf area per unit time. NAR is calculated from the following equation. $(W_2\text{-}W_1)$ $(\text{Log}_e\ L_2\text{-Log}_e L_1)/(t_2\text{-}t_1)$ $(L_2\text{-}L_1)$.Where, L_2 and L_1 are total leaf area at times t_1 and t_2 respectively; W_1 and W_2 are total dry weights at time t_1 and t_2, respectively.

Net Energy (NE): Difference between metabolizable energy and heat increment; includes the amount of energy used either for maintenance only or for maintenance plus production.

Net Irrigation Requirement: The amount of irrigation water just required to bring the soil moisture content in the root zone depth of the crops to field capacity i:e.it is the difference between the field capacity and the moisture content in the root zone before application of the irrigation water. NIR could be calculated using I following relationship. NIR= $(M_{fci}$ - $M_{ci})$ x $A_i.D_i/100$. Where, NIR= net amount of water (cm) to be applied during an irrigation; Mfci= field capacity moisture content in the Ith layer of the soil (% w/W); *Mbi*= moisture content before irrigation in the lth layer of the soil; Ai=. bulk density of the soil in the lth layer of the soil; Di= depth of the Ith soil layer (cm) within the root-zone; n= number of soil layers in the root zone D.

Netutral Soil: A soil that is neither significantly acidic nor alkaline having pH between 6.6 and 7.3.

Nick: In hybrid seed production, the parents are said to hick when they produce higher yield as a result of synchronized flowering.

Nitrogen Assimilation: The incorporation of nitrogen compounds into cell substances by living organisms.

No-tillage Planting: Direct planting in essentially unprepared seed beds solubility and high oil solubility.

Normal Solution (N): When gram equivalent weight of a compound is dissolved in one litre of water the solution is one normal solution.

Normal Year: A year during which the precipitation, stream flow, or other phenomenon approximate the average for a long period of record.

Noxious Weed: A weed arbitrarily defined by laws as being especially undesirable, troublesome, and difficult to control. Definition will vary according to legal interpretations.

Nurse Crop: 1. A companion crop which nourishes the main crop by way of nitrogen fixation and or adding the organic matter into the soil. e.g. cowpea intercropped with cereals or new plantations of fruit trees. 2. A physical method of weed control. They are usually quicker to germinate and are rapid in early growth.

Nursery: Places or piece of ground where seedling are raised for transplanting to the field.

Nut: A one-seeded, dry, indehiscent fruit, with a hard or bony pericarp, derived from a simple or compound ovary and characteristically, rish in oil e.g. acorns, chestnuts or hazelnuts.

Nutritive Value Index (NVI) : Intake of feed DM x digestibility of DM x efficiency of utilization of digested nutrients.

Oasis Effect: Effect of the dry fallow surrounds on the microclimate of a relatively small acreage of land, where an air mass moving into an irrigated area sill give up sensible heat. For small field, this may result in a higher ET crop as compared to predicated ET crop using a climatic data collected inside the irrigated area; conversely ET crop predictions based on weather data collected outside the irrigated fields may over predict actual evapo-transpiration losses.

One Third Atmosphere Moisture Point: The percentage of moisture retained by a soil sample when placed on a porous plate and subjected to a pressure of 1/3rd atmospnere (pF=2.54). It is supposed to be a good estimate of field capacity and moisture equivalent. Sometimes called the upper limit of readily Available water.

Organic Farming: According to the U.S. Department of Agriculture "a production system which avoids or largely excludes the use of synthetically compounded fertilizers, pesticides, growth regulators and livestock feed additives. To the maximum extent feasible, organic farming systems rely upon crop rotations, crop residues, animal manures, legumes, green manures, off-farm organic waste, mechanical cultivation, mineral-bearing rocks and aspects of biological pest control to maintain soil productivity and tilth, to supply plant nutrients and to control insects, weeds, and other pests".

Organic nitrogen (Soil): Nitrogen in the form of organic compounds e.g. amino acids, amino sugar, purine and pyrimidine bases.

Organic Phosphorus: Phosphorus present as a constituent of an organic compound or a group of organic compounds, e.g. glcerophosphoric acid, inosital phosphoric acid, cytidylic acid.

Osmosis: The diffusion of water across a differentially permeable membrane from a region of greater water concentration to a region of lesser water concentration.

Osmotic Potential (psi_o): The amount of work that a unit quantity of water in an equilibrium soil water system is capable of doing when it moves to another equilibrium system identical in all respects except that there are no solutions. It is also termed solute potential is the portion of water potential that result/from the solute present in the soil system: psi_o = -pi. Where pi=osmotic pressure due to dissolved salts.

Osmotic Pressure: The pressure that must be exerted on a solution to prevent the passage of solvent molecules into it when the solvent and solution are separated by a semipermeable membrane. As the' osmotic pressure is a colligative property it is directly proportional to the molar concentration of the solute if the temperature remains constant; thus PI is proportional to n/V where PI is osmotic pressure and n/V is the concentration.

Outer Crop/Guard Crop/Border Crop: The crop which are grown around the field boundaries in narrow strips with twin objectives of protecting the main crop from stray cattle and producing livestock feed and /or seed are called out or guard crop e.g. *Sesbania* or *Leucaena* on boundaries of field/plantation crops and castor around spring planted sugarcane.

Oven Dry Soil: Soil is considered to be oven dry when it has reached equilibrium with the vapour pressure from an oven at 105°C. The oven dry condition corresponds to a relative humidity of approximately 0% of pF near 7.

Overseeding : It refer to seeding of a legume/grass into an existing pasture in. order to improve the productivity and quality of herbage.

Parallel Cropping: A type of multiple cropping in which two crops of dissimilar growth habit are grown simultaneously in such a way that they do not affect the performance of each other adversely. A short duration, quick-growing parallel crop is raised in space left in between the rows of long duration main crop, the former mature without exercising any cognisable competitive influence on the latter.

Parasitic Weed: 1. Weed that partially or fully depends for its growth on some nutrient drawn from its host plant. 2. Weeds that attack other plants by making an organic connection with the conducting tissues of the host and deriving part or all of their food from the host. They either have no connection with soil or quickly lose it and are therefore difficult to control, e.g. mistletose, orobanche, birdvine, dodder, which weed, etc.

Parasitism: A relationship in which a heterotrophic organism obtains its food from the living tissues of another organism with no recompense to the host for example orobanche on tabacco.

Parboiling (Rice): Quit often the paddy is subjected to parboiling treatment before milling and it involves soaking the paddy in water, steaming it in hot water and drying it in the sun. Parboiling makes hust-removal easier, toughens the grain which result in reduced brakeage during milling and polishing and improves the keeping quality of rice. Also parboiled rice retains more nutrients while processing and cooking.

Parent Material: The horizon of weathered rock or partly weathered soil material from which the soil is formed. Horizon 'C' of the soil profile.

Passive Absorption of Water: The absorption of water by the roots as a result of forces originating in the leaves such as transpiration pull and does not require any expenditure of metabolic energy.

Pasture land: An area of land covered with grass or other herbaceous forage plants used for grazing animals.

Pasture Management: The application of practices to keep pasture plants growing actively over as long a period as possible so that they will provide palatable feed of. high nutritive value; to encourage the growth of desirable grasses and legumes, while crowding out weeds, bushes and inferior grasses.

Pedology: The science dealing with the laws of origin, formation and geographic distribution of the soil as a body in the nature.

Per cent Germination: the proportion of the number of seeds which have produced seedling classified as normal under the conditions and period specified. Per cent germination = No. of seeds germinated. Total no. of seeds under test x 100.

Percolation: A type of laminar flow occurring in interconnected openings of saturated granular material under hydraulic gradients commonly developed underground.

Permanent Crop Land Area: Land cultivated with crops that need not be replanted after each harvest, such as rubber, coffee, cocoa, including land under shrubs, fruit trees, nut trees and vines but excluding land under trees grown for food or timber.

Permanent Cropping System: Agricultural system in which crop plants are grown continuously on same plots of land, either several crops in rotation or single crop- for several years.

Permanence Pasture: A pasture of perennial or self-seeding annual plants kept for grazing indefinitely.

Permanent Wilting Percentage: The moisture percentage of soil at which plants wilt and fail to overcome wilting even though placed in a saturated atmosphere. Usually determined with dwarf sunflowers as the test plant.

Permanent Wilting: A wilting condition of the plant due to loss of water. The plants would not recover from wilted condition during the night and they require watering for their recovery from wilted condition and region of turgor.

Permeability: The property of a porous medium to transmit fluids. It is a broad term and can be further specified as hydraulic conductivity and intrinsic permeability. 'K' in Darcy's law and represents the rate of flow of water in porous medium under a unit hydraulic gradient. It changes with the quality and quantity of water but the value of intrinsic permeability, the property of porous medium remains constant for any fluid. they come to the top. Also called RAHAT.

Persistence: 1. Ability to withstand extremes of climatic, adaphic, and biotic stresses in nature. 2. Persistence is the perennation of the original plants, and maintain one of the species by vegetative spread or by regeneration from self-sown seed.

Persistent Herbicide: A herbicide which, when applied at the recommended rate will harm susceptible crop(s) planted in normal ortaition after harvesting the treated crop or which interferes with regrowth of native vegetation in non-crop sites.

Phenology: The study of the timing of recurring biological events, the causes of their timing with regard to biotic and abiotic factors and the inter-relation among the phases of the same or different species.

Photo-oxidation: When the light intensity exceeds a limit, leaves consume oxygen and use it in the oxidation of certain cell constituents. It is a chlorophyll catalysed process and more chlorophyll molecules become excited than possibly be utilized causing harmful side effect. It results in bleaching of chlorophyll and inactivation of important enzymes. Oxygen, light intensity and presence and absence of carotenoids and CO_2 concentration influence photo-oxidation.

Phototropism: Growth response of plants to variation in light intensities.

Phytochrome: A photo-morphogenic protein pigment in plants involved in photoperiodic response of plants. It is present in two inter, convertible forms, the red light absorbing form (pr) and other far-red light absorbing form (pfr).

Piezometer: It is a hollow pipe with opening at the bottom used to measure pressure of ground water at the point of entry.

Pitcher Farming: A practice in dry farming where crop is irrigated through small holes made in the bottom of earthen pitcher. The practice is generally used for wider spaced plants.

Placement (Fertilizer): Inserting, drilling or placing the fertilizer below the soil surface by means of any tool, implement or equipment at desired depth to supply plant nutrients to crop before sowing or in the standing crop.

Planking: One of the tillage operations carried out before sowing the crop for smoothening, microlevelling and to cover the seeds after sowing.

Plant Community: Assemblage of plants living together under the same environmental conditions.

Plant Ecology: The branch of ecology which deals with the study of plants in relationship with the environment.

Plant Growth Regulator: 1. A synthetic chemical other than a nutrient which can be used to modify plant growth. 2. Organic compounds occurring naturally in plants as well as synthetic other than nutrients which in small amount promote, inhibit or modify any physiological process in plants.

Plant Interference: The response of an individual plant or group of plants to its total environment as modified by the presence and or growth of other individuals or species of plants. It denotes total mechanism whereby a plant or a group of plants suppresses or otherwise modifies the growth of neighboring plants.

Plant nutrients: The elements taken in by the plant which are essential to its growth and used by it is elaboration of its food and tissues includes nutrients obtained from fertilizer ingredients.

Plant: An organism characterized chiefly by a holophytic mode of nutrition dependent upon the possession of chlorophyll. Some plants, however, are parasitic (see Parasitism) or Saprophytic. Most have cellulose cell walls, are sedentary, and have branching bodies. The distinction between plants and animals becomes blurred in some group (e.g., slime moulds, unicellular algae, protozoa), which are now often regarded as part of a third kingdom, the protista.

Plantation: Plantation weed an area in which trees have been planted especially for commercial purposes.

Plough Sole Placement: A method of fertilizer application in which fertilizer is placed in a continuous band in the. bottom of the furrow, in the process of ploughing each band is covered as the next furrow is turned.

Plumule: The major young bud of the embryo within a seed from which aerial portions of the plant develop. It usually occurs at the tip of the epicotyl, the part of the embryonic plant axis above the cotyledons.

Pluvial Rice Lands (Rainfed Uplands): Lands which are well drained without free ground water within the rooting zone of the rice plant and are generally unbunded.

Pod : An elongated seed vessel of a leguminous plant such as the Pea, splitting open on both sides when ripe.

Poisonous Weeds: These are responsible for the great loss sustained by the country every year, on account of the death of livestock. The plants may be harvested along with fodder crops or cut with grass and fed to cattle, unintentionally. But such neglect may lead to serious results. Sometimes cattle graze on poisonous plants in pasture land and death results only it the quantity eaten is sufficiently large.

Polarimeter: An instrument for measuring the polarization of light, epecially to determine the optical activity of a subsistence.

Polarity: It describes the electrical phenomenon of anion. The molecule or ion of the polar compounds have both electrically positive and negative regions. These polar substances are also called hydrophilic substances, have greater affinity with water and hence are soluble in water and other polar solvents.

Polishing: The removal of bran layer from brown rice by mechanical operation. It is also known as whitening or pearling.

Pollution: Contamination of natural environment by the addition to air or water of substances potentially toxic or otherwise harmful to man and animals for example, SO_2, CO, radio-active fall out, insecticides etc.

Pore Space: The empty space in soils not occupied by the soil solids. It is filled with air or water or both. It is expressed as the per cent of the total volume of the soil. It is calculated as shown below: % Pore space = 100 - per cent of space occupied by soil solids = 100 - (Apparent specific gravity) / (True specific gravity (i.e. 2.65)) x 100. Sandy soils with higher apparent specific gravity have less pore space that fine textured soils with lower apparent specific gravity. The compacted subsoil also have low pore space compared to surface soils.

Porosity: Porosity of soil (N) is defined as the percentage of soil volume not occupied by the soil particles. It is obtained from particle density (dp) and bulk density (bd) by the relationship, N = (1 - bd/dp) x 100.

Post Emergence Herbicides: A herbicide applied after the emergence of a crop or weed for the control of weeds.

Post Harvest Tillage: The cultivation of the land immediately after harvesting at physiological maturity to aid in the preparation of the land for the next crop.

Potable Water: Water which does not contain objectionable pollution, contamination, minerals, or infection, and is considered satisfactory for domestic consumption.

Potassium Fixation: The process of converting exchangeable or water-soluble potassium to a form not easily exchanged from the adsorption complex with a cation of a neutral salt solution.

Potassium Adsorption Ratio (PAR): A ratio for soil extracts and irrigation waters used to express the relative activity of potassium ions in exchange reactions with soil.

Potential Crop Yield: Yield of a crop obtained at research stations under most ideal environment.

Potential Evapotranspiration: 1. The amount of moisture that will be evaporated from the soil and transpired from vegetation if it were available. It combines moisture and heat factor in climate. 2. The evaporation that occurs when the ground is completely covered by actively growing vegetation and there is no limitation of the soil moisture.

Pre-emergence Herbicides: A herbicide that is applied prior to the emergence of a specific crop or weed.

Precipitation: Total amount of precipitation (rain, drizzle, snow, hail, fog, condensation, hoar frost and rime) expressed in depth of water which would cover a horizontal plane if there is no runoff, infiltration or evapo-transpiration; mm/day.

Preparatory Tillage: Various tillage operations carried out on the farm before the crops are sown, planted or transplanted to make a fine seedbed.

Pressure Plate Membrane Apparatus: The apparatus consists of ceramic porous pressure plates or membranes of high air entry values contained in airtight metallic chambers strong enough to withstand high pressure (15 bar or more).The apparatus enables the development of soil water characteristic curves in the higher range of metric potential (1 bar) which is not possible in suction plates.

Prey Crop (Smother Crop): The crop which is grown with the purpose of eliminating any undesirable plant through physical or its allopathic effect.

Primordial Stage: Initiation or first formation of ear head in quiet primary reproductive stage in plant body. The primordial formation occurs after the maximum tillering stage in rice, wheat etc.

Productivity (Plant): The capacity of a plant to produce yield or biomass per unit area per unit time.

Protected Spraying: A method of herbicide application. Non-selective herbicides can be employed to obtain selective weed control in distantly planted vegetables and ornamentals by covering the non-target plant with plastic or metallic covers.

Puddled Soil: Dense massive soil artificially compacted when wet and having no regular structure. The condition commonly results from the tillage of a clayey soil when it is wet.

Puddling: 1. The ploughing or working of land in standing water with country or desi plough or any other suitable implement. 2. The reduction in the apparent specific volume of a soil by doing mechanical work upon it.

Purity Test: It refers to the determination of the percentage purity of the item viz. pure seed obtained after discarding other crop seed, weed seed, inert matter based on examining the quantity specified under rules.

Quadratic Response: The output goes on increasing upto a certain level of input and then starts decreasing when input is increased further. In such cases the response in said to be quadratic and is given by: $Y=a+bx+cx^2$

Quality Seed: Good quality seeds, generally true to species or variety, have capacity for high germination, are free' from diseases, insects, weed seeds, insert matter, other crop seed and extraneous material.

Quarantine Laws: Are laws that enforce isolation of an area in which a severe weed has become established and to prevent the movement of the weed into an uninfected area.

Quarantine: Control of import and export of plants or parts of plants to prevent spread of diseases and pests.

Rabi Season: Season during which winter crops like wheat, barley mustard, gram, linseed, etc. are cultivated and usually it extends from September to April.

Radiant Flux: The amount of radiant energy emitted, transferred or received per unit time.

Radio Activity: Spontaneous nuclear transformation, the energy of the process being emitted in the form of alpha, beta or gamma rays,

Rain Factor (RF) : An expression of the humidity as a function of rainfall and temperature, as shown below: RF = (Mean annual rainfall (mm))/(Mean annual temperature °C.

Rainfall Index: The average rainfall intensity above which the volume of rainfall equals the volume of observed runoff. It represents the combined effects of interception and depression storage as well as of infiltration.

Rainfed Dry land: Areas that depend on rainfall for crop production, but are not flooded. The run-off and infiltration of rain water is so high that water does not accumulate on the land.

Rainfed Wetland: Areas where rainfed rice is grown on puddle soil in fields bounded by dikes capable of ponding water to about 30 cm. Water depth seldom exceeds 30 cm. These areas received no irrigation and non-rice crops may be grown before or after the wetland rice crop.

Rainy Day: A period of 24 hours in which 2.5 mm or more rain is recorded.

Range (Pasture): An extensive area of natural pasture land. If unfenced it is an open range.

Range Management: Management of a range of produce maximum forage for sustained use without jeopardinsing the other resources or uses of land.

Ratoon Cropping: the cultivation of an additional crop from the regrowth of stubbles of previous main crop after its harvest, thereby avoiding reseeding /replanting such as in sugarcane, sorghum, rice, fodder grasses etc.

Ratoon: The emergence of new tillers from the root/crown of a plant specially as in sugarcane, sorghum etc.

Reaper: A machine to cut grain crops.

Recession Farming (Diara Land Farming): It is a system in which crops are planted in flooded areas as the rainy season ends and water recedes. This system takes advantage of thoroughly saturated soil profile and also has the advantage of silt and nutrients left behind by flood water.

Reclamation (Land) : The operation or process of changing the condition or characteristics of land which cannot be utilized to its full potential otherwise viz., irrigation of land, drainage of swamp or waterlogged land, gypsum application to alkali soils, liming of acid soils etc.

Reclamation of Water: The treatment of sewage effluent or other waste so that it can be re-used directly, not necessarily for drinking. Two of the best known examples are at Lake Tahoe, California, and Windhoek, Namibia. In another semi-desert country, Israel, sewage effluent is regularly used for irrigation and other purposes. In Europe artificial recharge of groundwater with treated sewage effluent is form of reclamation, since the groundwater is later pumped out as raw water. Often the most expensive pollutant to be removed from sewage is chloride, because desalination is needed. Reclamation may involve removal of nitrogen, phosphorus, chlorides and trace organics, followed by conventional treatment for raw water.

Relative Crowding Coefficient (K): It is a measure of the relative dominance of one component crop over the other in an inter-cropping or mixed cropping system. The coefficient (k) is determined separately for each component crop e.g. for crop 'a' in association with b, the coefficient is as under: kab = Yab x Zba/(Yaa-Yab) x Zab.

Relative Density (RD): The population of a particular species expressed as a percentage of all the species in a habitat. Thus, RD = Density of a particular species/Total density of all the species x 100.

Relative Dominance: The actual area occupied by a species expressed as a percentage of total area occupied by all species in a given land area. (Total basal area of species 'a') / (Total basal area of all species) x 100.

Relative Growth Rate (RGR): An index of the amount of growing material per unit dry weight of plant per unit time. W_1 and W_2 are the total dry weights at time t_1 and t_2 respectively, the mean value of RGR for the time interval t_1 to t_2 is given as $Log_e W_2$-$log_e W_1$ / t_2-t_1. It is also called efficiency index (y) and can be expressed in g/g/day.

Relative Humidity (RH) : The ratio between the amount of water vapour in the air and the maximum which it could hold at the temperature concerned. The RH is a fraction, never more than 1. At 100 Per cent RH rain or dew is already falling or can soon be expected.

Relative Turgidity (RT): Ratio of turgidity of leaf tissue at a particular stage to that of when it is fully turgid. It is expressed as: RT = fresh leaf wt. - dry leaf wt./ fully turgid leaf wt. - dry leaf wt. x 100

Relative Yield Index: Is the ratio of mean yield of the crop in district to the mean all India (or national) yield.

Replenishment Irrigation: Artificial recharge of ground water by over-irrigation of crops, or off-season spreading of water on irrigated lands. Also called percolation irrigation.

Residual Herbicide: One which, as used, remain active in the soil for a short or longer period.

Residual Sodium Carbonate (RSC): The excess of carbonate plus bicarbonate over calcium plus magnesium in a water RSC = $(CO_3^{-2} + HCO_3^-) - (Ca^{++} + Mg^{++})$ Where the ionic concentrations are expressed in milliequivalent per liter.

Residual: 1. To have a continued killing effect over a period of time. 2. To continue over a period of time to have a killing effect on all or specific life forms.

Residue (Pesticides): That quantity of a substance especially of active pesticide, remaining in soil or plant.

Rest-Rotation Grazing: Intensive system of management whereby grazing is deferred on various parts of the range during succeeding years, allowing the deferred part complete rest for one year; two or more units are required.

Rhizobia: A group of symbiotically bacteria which are capable of fixing atmospheric nitrogen in the root nodules of legumes from which they receive their energy.

Rhizome: 1. A creeping horizontal swollen stem that acts as an underground storage organ and produces new shoots each season. Plants that grow from rhizomes include bearded iris. 2. An underground stem which spreads sideways from the plant and which possesses buds capable of producing aerial shoots and roots. 3. Horizontal, slender, underground root like stem that sends out roots and leafy shoots.

Rhizosphere: That portion of the soil in the immediate vicinity of plant roots in which the abundance and composition of the microbial population are influenced by the presence of roots.

Rice Glazing: Polished rice treated with talc (and in some cases with glucose).

Rice Milling: Involves the removal of hulls and bran from paddy rice. Milling ranges from hand pounding to modern milling equipments.

Ring-Barking: A method of killing trees by removing a ring of bark, down to and including a cambium near the bottom of the trunk.

Root Crop: Crop plants whose surplus or reserve foods are stored primarily in enlarged roots e.g. sweet potato.

Root Exudates: Exuded matter from roots or discharge of chemical(s) through roots.

Rosette: In many plants profuse leaf development with retarded internode growth to give bunchy or rosette appearance.

Rotational Grazing: Dividing a large pasture into several small paddocks and allowing the cattle to graze each, periodically perhaps every four weeks.

Rouguing : To remove weeds or off-type or diseased plants from a standing field crop.

Run-off: When spraying, the point at which a plant becomes saturated, and further liquid runs off on to the surrounding area.

Runner: A trailing stem that grows along the surface of the soil and takes root and forms new growth at the nodes.

Runoff Coefficient: 1. The ratio of the maximum ratio of runoff to the uniform rate of rainfall with a duration equaling or exceeding the time of concentration which produced this rate of runoff. 2. Used incorrectly for percentage runoff. 3. The ratio of the depth of runoff from the drainage basin to the depth of rainfall.

Runoff Plots: Small areas of land from which surface flow may be measured. Used for comparing different vegetative or mechanical treatments for runoff and soil loss

Salt Stress: This stress result from an excess of anyone of a large number of salts. Plants showing salts (sodium) resistance are called HALOPHYTES.

Salt Tolerance: Is the ability to tolerate high concentration of soluble salts in the root medium. It usually is expressed by reference to the salinity level that causes a certain decline in yield.

Sampling Unit: A defined quantity of material having a boundary which may be physical, for example a container, or hypothetical, for example a particular time or time interval in the case of a flow of material.

Sand Culture: A technique for the study of plant nutrition in which solid media (sand or crushed quartz) are used. The important advantage of solid media over liquid media is that roots are glowing in a natural medium and no means of support needs to be provided. Nutrient solutions are added to the solid culture by three different means (i) Slope culture involving pouring over the surface, (ii) Drip culture involving dripping on the surface, and (iii) Subirrigation forcing solution up from the bottom of the container. In all the three methods the added nutrient solutions pass out through an opening in the bottom of the container.

Satellite Weed: Weed that has become an integral part of a crop ecosystem.

Saturated Air: Moist air in a state of equilibrium with a plane surface of pure water or ice at the same temperature and pressure. In such a state the relative humidity is 100 per cent and the amount of water vapour is maximum for the given temperature.

Saturation Percentage: The moisture percentage of a saturated soil expressed on a dry-weight basis.

Scarification (Seed): Scratching of hard seed coat to remove the mechanical barrier to obtain uniform germination with better penetration of moisture into the seed.

Scouring or Pearling (Rice): Process of bran removal (locally known as polishing).

Season Bound Weeds: All weeds which grow in specific season of the year with disregard to the crop species cultivated. These weeds may be either summer annuals or winter annuals.

Seasonal Consumptive Use: The total amount of water used in evaporation and transpiration by a crop during the entire growing season. It is expressed as depth of water in cm or volume in hectare cm per hectare. Seasonal consumptive use values are required to evaluate and determine seasonal irrigation water supplies.

Seasonal Drought: A kind of drought which occurs during distinct annual periods of dry weather.

Seed: Sexually or vegetatively propagated planting materials which are used for seeding and planting and as such should be free from (pests and diseases) any infection and should give a good crop stand by good seeding.

Seed Bed: A well prepared land for sowing/planting.

Seed Control: Seed law enforcement which applies to all Seeds sold commercially. Correct labelling with respect to the kind of seed, germination, purity and other quality factors are ensured.

Seed Dressing: To chemically teat seeds before sowing to control any disease or pest attack in growing crops, particularly cereal.

Seed Hardening: It is a process of subjecting seeds before sowing to alternate cycles of wetting and drying to induce tolerance to drought.

Seed Rate: The quantity of seed required for sowing on a unit area of land.

Seed Treatment: The process of application of a thin coat of insecticide or fungicide over seeds or similar applications to prevent pest infestation.

Seed-Cum-Fertilizer Drill: An implement which drills seed and fertilizer simultaneously and uniformly in the same or different rows.

Seedling Stage: It includes the period from emergence till just before the appearance of first tiller.

Seepage: The slow movement of water through small cracks, pores, interstices, etc., in the surface of unsaturated material into or out of a body of surface or sub-surface water.

Selective Herbicide: 1. A chemical which kills or stunts some plants with little or no injury to others. 2. Refers to a herbicide that is more toxic to one plant than to another. When such a herbicide is applied to a mixture of plants, some may be killed and others may be affected only slightly or not at all. 3. An herbicide that will kill some plant species when applied to a mixed population but without serious damage to other species.

Selectivity Index: Ratio of ED_{50} of test crop to ED_{50} of associated weeds.

Semi-Parasitic Weed: Parasitic weeds like striga having green leaves which enable them to manufacture carbohydrates; solutions of salts are obtained from the host.

Seminal Root: A root arising from the base of hypocotyl, as in wheat.

Semi permanent Soil Sterillant: A chemical which makes the soil sterile for more than 4 months and less than 2 years.

Sequential Application: Two or more herbicide treatment applied at different times to the same crop.

Sessile: Without a stalk; sitting directly on its base.

Sexual Reproduction: Conjugation and fertilization of gametes resulting in spore and seed formation, respectively.

Shelling: Removal of grains from cobs or husk from paddy.

Shelter Belt (Wind Break): A wind barrier of living trees and shrubs established and maintained for protection of crop fields.

Shifting Cultivation: The practice of cultivating clearings scattered in the reservoir of natural vegetation (forest or grass-woodland) and of abandoning them as soon as the soil 'is exhausted; and this includes the practice of shifting homesteads in order to follow the cultivator's search for new fertile land.

Side Dressing: The application of commercial fertilizer along the side of a row or around a plant.

Silique: A dry, dehiscent fruit, that is two celled. The valves split from bottom up and leave the placentae with the false partition (replum) between them. Fruit narrow and longer than wide; characteristic of many species in the mustard family (Brassicaceae).

Silo: A semi-airtight structure designed for use in the production and storage of silage.

Siltation: The deposition or accumulation of unconsolidated or loose sedimentary material whose constituent rock particles are finer than grains of sand and larger than clay. Siltation can result in the silting up of reservoirs and streams and in the frequent clogging of irrigation channels.

Silting: The deposition of water-borne sediments in lakes, reservoirs, stream channels or overflow areas due to decreased velocity of flow. Silting provide added medium for the growth of rooted aquatic plants, and place their seedlings closer to sunlight which encourage the growth of aquatic weeds.

Silviculture: is the art and science of cultivating forest crops as distinguished from silvics which deals with the study of the life history and general characteristic of forest trees of crops.

Silvipasture : The practice of growing forage crops in association with forest crops simultaneously on the same land at the sametime. Under the silvipastoral system in intervening spaces between the two rows of trees are utilized for growing nutritions grasses/ grass legume mixture instead of allowing weeds to grow in its place. It is an integrated approach of management of land, water and plant resources to fix higher solar energy in less productive area.

Simpson's Index (Sol.): S.I. is a measure of concentration of dominance, can be used to determine the degree of diversity in a given community type. S.I. = $S(y/n)^2$ Y= Importance Value (I.V.). N= The sum of I.V.'s for all species in the sample.

Single Grain Soil: A structure less soil in which each particle exists separately as in sand dunes.

Sink: The term usually applied to refer to the plant parts that are involved in utilization and storage of photosynthates for example grains fruits etc.

Skip Row Planting: When a line is left unsown in regular sowing.

Slick Spots: Wet spots with high exchangeable sodium occurring generally in salt affected areas. .

Sludge: The semi-solid part of sewage water that has been sedimented or acted upon by bacteria.

Slurry: Semi liquid effluent from livestock, consisting of urine and faeces, possibly diluted with water.

Small Farmer: A farmer who is potentially viable. He is having just enough land which would enable him to become a surplus producer with the adoption of new technology. For adopting new technology he requires special assistance for procuring necessary inputs including credit.

Social Forestry: It is a programme of forestry development and conservation under various agro-climatic and soil conditions through(a) mixed plantation on waste lands and panchayat lands and (b) reforestation of degraded forests and raising shelter belts.

Sod Strip: A narrow strip of grass used for checking erosion in waterways.

Sod: Top few centimeters of soil permeated by and held together with grass roots or grass-legume roots.

Sodium Absorption Ratio (SAR): A pointer to the effect of the sodium in a water on the soil and crops: $SAR = Na / \{O.5 (Ca + Mg)\}^{1/2}$. Where Na, Ca and Mg are the concentrations in milliequivalents per litre of sodium, calcium. and magnesium in the irrigation water. For crop irrigation the SAR should be less than6, while an SAR above 15 implies that the water is unsuitable.

Soft Water: Water with little or no carbonate hardness or non-carbonate hardness. In Britain it usually originates in the hard rock areas of the west and north, especially in Wales and Scotland, or from peaty moorland, but not from chalk, nor from limestone formation. Soft water may be corrosive, as in plumbo-solvency.

Soil Amendment: Any substance that is added to the soil for the purpose of improving its physical or chemical character, enhancing soil productivity or promoting the growth of crops but excluding commercial fertilizers and organic manure.

Soil Conditioners. Are linear flexible organic polymers similar to soil polysaccharide.

Soil conservation: A combination of all management and land use methods which safeguard the soil against depletion or deterioration by natural or by man-induced factors.

Soil Crusting: Is defined as hardening of surface soil of 2-5cm thickness. It differs from soil compaction which refers to hardening of soil of more than 5 cm depth. Soil crusting creates soil capping which leads to considerable runoff of rain water. Soil crusting after seeding resulting form soil dispersing rain and sudden drying results in improper and poor emergency of germinating seedling.

Soil Degradation: is a process of changing a soil from one type to another more highly leached one; particularly bringing about replacement of sodium by hydrogen by leaching a saline or alkali-soil.

Soil Erodibility Factor: The soil erodibility factor indicates the inherent erodibility of a soil. It given an indication of the soil loss from a unit plot 22 mm long with a 9% slope and continuous fallow culture.

Soil Moisture Deficit: The amount of water needed to return the soil moisture status to the field capacity.

Soil Moisture Stress: It is the sum of soil moisture tension and the osmotic pressure of the soil solution.

Soil Mulches: Are mulches created by stirring the surface soil until it is loose and open. It reduces evaporation from the soil surface.

Soil Penetrometer: Is an instrument for recording the force required to push a blunt-pointed probe into the soil. Other conditions being equal, soil with good structure for most agricultural plants are penetrated more easily than soils with poor structure.

Soil Sickness: Condition of soil when the productivity is reduced inspite of available nutrients.

Soil Sterilant : 1. A herbicide that prevents the growth of plants when present in the soil. The soil sterilization effects may be temporary or relatively permanent. 2. A material which renders the soil incapable of supporting plant growth. Sterilization may be temporary or relatively permanent.

Soil Texture: Characterization of soil in respect of its particles size and distribution.

Soil Water: The suspended water in the uppermost belt of soil of the zone of aeration and lying near enough to the surface to be discharged into the atmosphere by the transpiration of plants or by evaporation from the soil. Includes hygroscopic, capillary and non capillary water.

Sole Crops: A crop grown in .pure stand at optimum population (spacing).

Soil Moisture Tension: It is the force with which the water is held by soil against gravity. It is generally express in atmospheres or in mm of mercury.

Soluble Sodium Percentage (SSP): A term used in connection with irrigation waters and soil extracts to indicate that proportion of sodium ions in solution in relation to the total cation concentration. It may be calculated by the formula; SSP = {Soluble Na concentration (me/l)} / {Total cation (me/l)}.

Solution: 1. A liquid which is physically homogeneous mixture of two or more substances. 2. A physically homogenous mixture of solute and the solvent.

Solvent: A substance (usually water) used to alter the form of a compound; i.e. cause it to flow freely.

Sorghum Injury: Poisoning of livestock resulting from feeding of young sorghum plants containing Hydrocyanic acid (HCN). HCN concentration is more at seedling stage.

Spatial Arrangement: It is defined as the pattern of distribution of plants over the ground, which determines the shape of the area available to the individual plant.

Specialized Cropping. It is a cropping plan in which a single crop contributes 50% or more of the total productivity or monetary receipt (comparable equivalent) in one year.

Specific leaf area: The leaf area per unit leaf weight expressed as cm^2g^{-1} and is given by: Specific leaf area = area of the leaf (cm^2) / weight of the leaf (g).

Split Application: The total amount of a herbicide divided between two or more applications made at different times to the same crop.

Spot Application: A method of herbicidal application in which the chemical is directed to patches of weeds.

Spot Treatment: 1. Application of sprays to localized or restricted areas as differentiated from overall, broadcast, or complete coverage. Sport treatment is usually done on small areas of serious weed infestation to kill it and to prevent its spread. 2. Application of herbicide to small patches of weeds, leaving the weed-free gaps untreated.

Spray Drift: 1. The dispersal of chemical sprays by the wind on the neighboring areas where the chemical may cause damage. 2. Movement of airborne spray droplets beyond the target area. 3. The movement of airborne spray particles. The amount of spray drift depends upon the size of the droplets, amount of wind, and the height above the ground that the spray is released.

Starter Fertilizers: This fertilizer is complementary to the main fertilizer, applied at sowing or planting, in a very soluble form, in order to give extra nutrition to the plant during its young stage. This initial stimulus improves the early growth and eventually the yield.

Starter Solutions: Solutions used to provide readily available nutrients for early growth of a transplant or a seedling as it emerges from the seed. Some growers regard starter solutions as a valuable aid in early growth of seedlings prior to the time that roots come in contact with the main body of fertilizer placed in the soil.

Stolon: A creeping stem at or just below soil level that can root at the tip to produce a new plant. A plant that has stolons is said to be stoloniferous.

Strip Cropping: The practice of growing crops by a systematic arrangement of strips or bands. Commonly cultivated crops and sod crops are alternated in strips to protect the soil and vegetation against running water or wind. The alternate strips are laid out approximately right angles to the prevailing direction of the wind where soil blowing is a hazard.

Strip Farming: Farming on land where large fields and long slopes are protected from water erosion by ploughing on the contour and planting alternate strips to grass and row crops. See contour farming.

Stubble Mulch: A mulch consisting of the stubble and other crops residues left in and on the surface of the soil as protective cover during the preparation of the seedbed and during the best part of the growing of the succeeding crop.

Suberization: The formation of a complex wax like substance in the cell walls of wounded or cut plant tissue to reduce water loss and protect the area.

Submerged Aquatics: These aquatic weeds are normally attached to the bottom of a ditch and grow entirely beneath the surface of the water. They cause the most trouble from the standpoint of reduced water flow in irrigation ditches. The most common submerged aquatics are sago pondweed (also called horsetail moss), stonework, water milfoil, cabomba, and waterweed (Elodea).

Sub soiling: Breaking the compact subsoils without inverting them with a special narrow cultivator, showel or chisel at depth usually from 30-60 cm from the surface.

Substitution Cropping: It is -substitution of an existing inefficient crop with an identified efficient crop in a zone or region.

Succession: An orderly sequence of different plant communities in an area. One community replaces another until the climax in vegetation is reached.

Sucker: A shoot growing from a stem or root at or below ground level, often some distance from the main stem.

Summer Fallow: land that is uncropped in summer and kept cultivated throughout a growing season. The most important function of summer fallow is storage of moisture in the soil besides promoting nitrification and controlling weeds.

Sunshine Recorder (Heliograph): Device which records the time interval during which solar radiation reaches the earth's surface with sufficient intensity.

Supplemental Irrigation: The watering of crops in regions where normal rainfall ordinarily supplies most of the moisture. It is provided when crop is facing moisture stress due to dry spells.

Support Crops: Certain fast growing crops which work as support to vine crops e.g. castor.

Surface Application: Soil active herbicides are commonly applied to the surface of the soil where they may be either left undisturbed or incorporated into the soil physically.

Surface Soil: The upper 12 to 20 cm of the soil, or in arable soils, the depth commonly stirred by the plough.

Suspension: A liquid or gas in which very fine solid particles are dispersed, but not dissolved.

Symbiotic Nitrogen Fixation: Fixation of atmospheric nitrogen by leguminous plants with the help of root nodule bacteria (*Rhizobium* spp.).

Symplast: The living continuum in plants through interconnecting protoplasm.

Sympodial, Sympodium : A branching growth pattern in which the main axis is formed by a series of successive secondary axes, each of which represent one fork of a dichotomy.

Synergism: 1. Cooperative action of different chemicals such that the total effect is greater than the sum of their independent effects. Opposite antagonism. 2. Complementary action of two or more chemicals such that the combined effect is greater than the sum of their independent effects.

Synergistic Effect: An interaction term to show that the total effect of a combination is greater or more prolonged than the sum of the effects of the two taken independently.

Systemic Herbicide: A compound which is translocated readily within the plant and has an effect throughout the entire plant system.

Systemic or Translocated Herbicide: Herbicide capable of moving within the plant to exert effects throughout the entire plant system irrespective of its place of entry e.g. 2, 4-D and Atrazine.

Systems Approach: Studying a system as an entity made up of all its components and their interrelationships, together with relationships between the system and its environment.

Tassel (Maize): The flower cluster at the tip of a com plant comprised of pollen-bearing flowers i.e. the staminate inflorescence of maize.

Temporary Wilting: Wilting condition of the plant due to loss of water which would recover from wilted condition during the night without addition of any water.

Terrace Cultivation: The growing of crops on level steps or terraces that have been constructed on hillsides. Banks or walls are built to hold the soil in place and, where paddy is grown, to enable water to stand on the field. Terrace cultivation takes place where land is scarce.

Terrace: 1. A flat, level or nearly level, narrow, area of land bordering a river to lake, bounded on at least one side by a definite steep slope rising upwards from it, and on the other sides by downward slopes. 2. A low embankment or ridge of earth constructed across a slope to control surface runoff and minimize soil erosion. 3. Sloping ground cut into a succession of benches and steep inclines for purposes of cultivation. Often the inclines are made quite steep and are protected by riprap or retaining walls are substituted, thus giving greater areas for cultivation. 4. Areas bordered by low broad ridges constructed on cultivated land of such alignment, height, and spacing as to conform to the topography and to permit travel by cultivating and harvesting machinery, the object being to prevent loss of soil by erosion.

Terrestrial Weeds: Weeds which normally start and complete their life cycle on land.

Test weight: It denotes the weight, of thousand grains (normally) of field crops. In some crops hundred seed weight is also considered.

Thallophytes: Are aquatic weeds without distinct leaves, roots, and stems. There is no gamatic union to lead to embryo development.

Thermo-hygrograph: Instrument resulting from the combination of a Hygrograph and Thermograph. It furnishes on the same chart simultaneously a continuous recording of atmospheric temperature and humidity.

Thermograph: Apparatus used to give a continuous record as a graphical chart of temperature with time.

Thermoperiodism: The term thermoperiodicity or thermoperiodism designates the effects of alternations of temperature between the day and night periods upon the reactions of plants.

Thinning: Removing of extra plants from thickly populated crop stand with an idea of maintaining optium plant population.

Threshold Yield: The starting point of yield response to a given nutrient under specified conditions. On a graph of yield response to addition of a nutrient will be the origin of the response function. Experimentally the threshold yield will be established by the treatment in which the nutrient in question is deleted from an otherwise complete or adequate fertilizer application.

Tissue culture: The growth of detached pieces of tissue in nutritive fluids under aseptic conditions.

Tolerant: 1. Not susceptible to (injured by) a pesticide application. 2. Capable of withstanding effects. For example, grasses are tolerant of 2-4-D to the extent that it can be used selectively to control broad-leaved weeds in cereals, pastures, and turf.

Top Dressing: The application of manure or commercial fertilizer to a crop that is already established.

Top-Soil: The upper layer of dark fertile soil in which plants grow. Below this lies the sub soil, which is lighter in colour, lacks organic matter and is often low in nutrients.

Topping (Tobacco): The removal of the terminal bud with or without some of the small top leaves just before or after the emergence of a flower bud.

Total Weed Control: 1. That form of weed control which involves killing all plant cover. Also known as non selective weed control. 2. Use of herbicide to kill all vegetation on un cropped land in situations where selectivity is not required.

Toxicity: 1. is another name for poisoning power some chemicals have a high toxicity and some have a low toxicity. Toxicity is further described depending as how long the poisoning takes to show effect. (i) Acute toxicity- The poisoning which results from a single dose or exposure to the chemical. Acute poisoning usually arises from an accident or carelessness, unless the poisoning is suicidal. Substances can be acutely toxic by way of inhalation, taking through the mouth or skin absorption. An Antidote is applied and should be on hand to offset this type of poisoning. (ii) Chronic toxicity- The poisoning which results from the intake or absorption of repeated small doses. The symptoms are vague and un-defined, likely to change from person to person, and extend over a period of time. The deep seated effects of this type of cumulative poisoning are often much more damaging than those of acute poisoning. The poison may be taken in as outlined above under acute toxicity or may be taken in because of continuous exposure to pesticides and/or consumption of food stuffs carrying excessive residues of such chemicals. 2. The ability of the chemical to alter the biological environment.

Toxin: A poisonous secretion of a plant or an organism.

Tracer Element: The isotope of an element which may be incorporated into a sample to trace the course of the element, alone or in combination, through chemical, biological or physical processes.

Translocated Herbicide: A herbicide which moves within plant from the point of entry to other plant parts (sinks).

Transpiration: Loss as water vapour through the leaves of plants, of water brought up from the soil through the roots. It increases with wind, sunshine and temperature and as the plant increases its area of foliage. Transpiration varies from plant to plant. Soil moisture does not greatly affect it until the wilting point of the soil is reached. See evapotranspiration.

Trap Crop: A crop employed to induce germination of specific parasitic weed without succumbing to it.

Triple Cropping: Growing three successful crops in an agriculture year on the same piece of land e.g. rice-rice-rice, rice-wheat green gram, pulse-maize-wheat etc. its intensity being 300 per cent.

Truck Crops: Crops which yield in tonnes and grown for distant market requiring heavy transport.

Tuber Crop: Crops possessing enlarged underground reproductive stems, roots e.g. potato, cocoyam, sweet potato etc.

Turbidity: The degree of opaqueness of the water due to the amount of fine (particles) in suspension. The particles that cause turbidity may also determine apparent colour. Carp, suckers, and sturgeon commonly extract food from the bottom deposits of lakes. Their feeding involves sucking, digging, and rooting which creates localized turbidity, and this in turn may adversely affect the growth of certain aquatic plants. Their feeding action may also be considered as a form of cultivation that occurs in shallow water that is otherwise protected from wave and current action.

Turgor Potential (Ψ_p): Part of total plant water potential arising from the forces exerted on the walls from the water attracted to the cell by the solutes and solids in the protoplast. The turgor potential is also termed the turgor pressure, wall pressure, hydrostatic pressure or pressure potential.

Turgor Pressure (TP): The pressure set up inside a plant cell due to the hydrostatic pressure of the vacuole contents pressing against the cell wall.

Ultimate Wilting Point: The moisture content of the soil at which all the levels of plants growing in that soil are completely wilted and will not recover in an approximately saturated atmosphere without addition of water to the soil. The lower end of the wilting range.

Up Land: Cultivable land situated on high ground or land which has perfect natural drainage.

Variety: A sub-division or a group of plants within a species which is characterized by growth, plant fruit, seed or other characters by which it can be differentiated from seeds of same kind.

Vegetative Reproduction: 1. Growth of specific, detached vegetative organs of plants into new individuals. 2. A means of weed persistence. Rhizomes, rootstocks and other vegetative propagules of perennial weeds are located as deep as one m in soil. This enables the weeds to persist for years even after their aerial growth is destroyed.

Vernalizatoin : Low temperature treatment of seeds and seedlings, which has the effect of artificially 'aging' the plant and reducing the time between sowing and flowering.

Viable: Able to survive and develop; as with viable seeds that remain alive and are able to germinate.

Volatile Herbicide: A herbicide having a sufficiently high vapour pressures that when applied at normal rates and normal temperatures, its vapoures may cause serious irreversible injury to desirable plants away from the site of application. Note: Vapour injury from spray drift are often difficult to distinguish.

Water Balance: The balance between demand and supply of the water for a given place and time, also called inflow-out flow, may be represented by the following equation: Water Balance = ET + Surface runoff + Sub-surface drainage + Change in soil water content.

Water Conservation: The physical control, protection management, and use of water resources in such a way as to maintain crop, grazing and forest lands vegetal cover, wildlife habitat for maximum sustained benefits to people, agriculture, industry, commerce, and other segments of the national economy. **Water Deficit:** Shortage of water or inadequate water supply. The amount of water needed to return the soil moisture status to the field capacity (FC).

Water Harvesting: 1. The practice of collecting water from an area treated to increase run off from rainfall and snow-melt. 2. The process of collecting natural precipitation from prepared water-sheds for beneficial use. 3. The collection of runoff rain water from treated or untreated land surfaces and storing it in open water reservoirs or in the soil itself.

Water Holding Capacity: The weight of water held by a given quantity of absolutely dry soil when saturated.

Water logging: The saturation of soil with water, causing the water table of soil to raise high enough to expel normal soil gases and interfere with plant growth or cultivation. It is an major irrigation problem associated with poor drainage. It causes soil degradation.

Water Potential: 1. Refers to the difference between the activity of water molecules in pure distilled water at atmospheric pressure at 30°C (standard conditions) and the activity of water molecules in any other system. The activity of these water molecules may be greater (positive) or less (negative) than the activity of the water molecules under standard conditions. 2. The amount by which the chemical potential of water in system is less than that of the pure water under standard conditions. It enables the prediction of the direction of osmosis and the relative rate of diffusion.

Water Regimes: Regime means level; different levels of water (standing or flooding) depth applied in irrigation experiments. Such as 2.5, 5, 7.5, 10 cm etc. standing water to saturation or field capacity (FC) to rice crop etc.

Water Saturation Deficit (WSD) : Parameter for measuring water deficits in plants. It is calculated from the leaves as shown below: WSD = (Fully turgid wt. - Fresh wt.) / (Fully turgid wt. - dry wt.) x 100.

Water Shed: The total run-off from an area which supplies water to a river or lake. A catchment area; a drainage basin.

Water Table: The upper boundary for ground water. The upper surface of the locus of points at which the pressure in the ground water is equal to atmospheric pressure.

Waterlogged: 1. Filled or saturated with water, or so filled with water as to be heavy or unmanageable, like a completely saturated log 2. A condition of land where the ground water stands at a level that is detrimental to plant growth. It may result from excessive irrigation or seepage, coupled with inadequate drainage.

Watershed Management: The planned use of watershed lands in accordance with redetermined objectives, such as the control of erosion, stream flow, sedimentation, and the improvement of vegetative cover and other related resources

Weed Competition Index: 1. May be defined as the reduction in crop yield due to presence of weeds in comparison to weed-free plot. 2. Gain in crop yield due to weeding as percentage of yield of weed-free crop.

Weed Control Efficiency (WCE): It denotes the efficiency of the applied herbicide or a herbicidal treatment for comparison purpose by using the following formula: WCE = {(Weed population in control plot) - (Weed population in treated plot)} / {Weed population in control plot}. or WCE = {(Dry mater of weed in control plot) - (Dry mater of weeds in treated plot) / (Dry matter of weeds in control-plot).

Weed Count: A measure for quantitative assessment of weeding treatments. Weeds may be counted by placing a quadrant in the plot at random and counting the number of plants inside the quadrant.

Weed index: It is an index expressing the reduction in yield due to the presence of weeds in comparison with weed free situation.

Weed Laws: Are laws enforced to reduce the spread of weed species and to provide well adapted high quality crop seeds.

Weed Legislation: The law made by the appropriate authority regarding the prevention, restriction, introduction, movement and eradication of weeds.

Weed spectrum: Refers to different kinds of weed species a particular herbicide will kill or inhibit.

Weighting Lysimeter: A method of determining consumptive use of crops by means of tanks which are weighed. These tanks are placed in surrounding of natural growth of the same species that the planted in the tank.

Weir: 1. A dam in a watercourse to stop the flow and raise the level. 2. A dam in a waterway or conduit used to control the water level or the flow. 3. A notch over which liquids flow and which is used to measure the rate of flow. A dam across a stream for diverting or measuring the flow. Note: The essential difference between an orifice and a weir is implicit in the expression; water flows through an orifice but over a weir, Orifice usually refers to comparatively small devices whereas weir includes both small and large structures. Further, the stream of water which is known as jet when issuing from an orifice becomes the nappe, sheet or vein when flowing over a weir.

Wet Bulb Thermometer: It is one of the two thermometers of a psychrometer whose bulb remains wet with thin film of distilled water and indicates wet-bulb temperature.

Wet Climate: A climate in which precipitation effectiveness is such that rain-forest vegetation prevails.

Wet Crop: Crop which depends on irrigation for its growth.

Wet Land: Pertaining to soils flooded for at least several weeks each year, or to crop grown in such soils.

Wilt: The collapse of leaves due to unfavorable water relations. It may be due to excessive transpiration, blocking of the xylem elements, pathogens or parasites, a vascular disease that interrupts the normal uptake; or a toxin produced by an organism.

Wilting Coefficient (WC): The water content of the soil at the time when the leaves of the plant under study first undergo permanent reduction in water supply. By a permanent reduction is meant a degree of wilting from which the leaves cannot recover in an approximately saturated atmosphere unless water is added to the soil.

Wilting Point: The minimum quantity of water in a given soil necessary to maintain plant growth. When the quantity of moisture falls below this, the leaves begin to droop and shrivel up. In any given soil the quantity is practically constant for all plants, but it increases with a decrease in the size of soil particles. The moisture content of a given soil at which the force holding water to the soil particles equals the maximum water absorbing force of the plant roots.

Wilting Range: The range in moisture content of the soil between the permanent wilting percentage and the ultimate wilting point.

Wind Brakes (Shelter Belts): A strip of trees or shrubs of crop plants serving to reduce the force of wind and provide a protective shelter against wind.

Wind Energy: It is also a form of solar energy since winds are caused by variations in the amount of heat that sun sends to different parts of the earth. Electricity is produced when a wind will catch the wind and revolves, rotating a turbine which powers electric generator.

Wind: Wind refers to air in motion, is best expressed in Km/hr. it is measured by anemometer.

Wind-Vane: Device used to indicate the directions from which the wind is blowing.

Xerophyte: A plant which tolerates dryness, particularly dryness of the soil, often such plants have poor development of the surface and a great root development, which enables them to recover water from the deep layers.

Yield Attributes: Are components which finally make up or control yield of any crop for example yield components of cereal grain yield are grain weight per grain, grain number/ear, ear number/ plant and plant number/unit area.

Yield: Harvested produce obtained from crop grown in a unit area of land usually expressed as quintals or tonnes per hectare.

Yield Potential: Full production capability of crop.

Yield Stability: The maintenance of yield at desired level over a period of time. Its index is measured by inverse of coefficient of variation, regression between yield and environmental index and disaster level estimated by probability of a system.

Zero Tillage: A term used when crops are grown only with the tillage of planting (no seed-bed preparation), usually Possible when all weeds are controlled by more than one application of various weedicides. Also called plough less farming, crop residue planting or slot planting or minimum tillage.

❑❑❑

2 Entomology

Abbott's formula: Mathematical formula used to correct for mortality in the untreated check e.g. Corrected % mortality = % Mortality in treatment - % Mortality in check x 100 / (100 - % Mortality in check).

Abdomen: The third division of the insect's body; it is usually composed of ten or eleven segments and bears functional legs in the adult stage.

Abduction: Drawing back or retraction.

Abductor muscle: Any muscle that opens out or extend in appendage or draws it away from the body.

Abiotic: Nonliving, pertaining to physico-chemical factors such as light, heat etc.

Aboral: Pertaining to the surface of the body opposite to that which bears the mouth.

Absolute temperature: It is defined as T = C+273.15 where C is temperature in degree Celsius. Zero degree T or absolute zero is -273.15°C,

Absorbent (formulation): A substance capable of absorbing other substances usually on the basis of the wetting ability.

Acaricide: Agent destructive to acarids (ticks and mites) and used in the treatment of mange. Also called MITICIDE.

Acarology: The study of ticks and mites.

Acceptable daily intake: The amount of a toxicant that, if ingested daily with food is devoid of any adverse effect on the subject. It is defined as the daily exposure level of a toxicant in mg kg^{-1} body weight which during entire life span appears to be without appreciable risk (devoid of any injury or harm), on the basis of various facts taken at that time. It is generally determined by dividing the no effect level of the toxicant by a safety factor (generally 100).

Accessory glands: In the female, a pair of glands opening primarily on the venter of the ninth abdominal segment, secreting an adhesive substance or material forming a covering or a case (ootheca) for the eggs; in the male, mucous glands opening into the ejaculatory duct.

Accessory vein: A secondary longitudinal vein in an insect wing.

Acclimatization: Adjustment to environmental change on the part of the individual; the physiological adjustment or increased tolerance shown by an organism to environmental change; nonstandard variant of acclimatization.

Accumulation and elimination (of Toxicant): Accumulation is the total amount of toxicant accumulated inside the body (tissue) following a regular up-take/ingestion of a certain amount daily. Elimination is the amount of toxicant degraded or metabolized by the system. So the net amount left after elimination in various components (liver, fat, blood etc.) of the organism is accumulation.

Acephate: is an organophosphate foliar insecticide of moderate persistence with residual system activity of about 10-15 days at the recommended use rate.

Acetylcholine: A chemical conductor of a nervous impulse, formed at the ends of nerves, to conduct the impulse over the gap between nerves or between nerves and muscles or glands. After the response, ACh is destroyed by the enzyme cholinesterase.

Acone eye: A type of insect compound eye in which the ommnidium is neither a crystalline nor a liquid cone in place of which there is a group of elongated transparent cells.

Acron: In the arthropod embryo, the head region anterior to the tritocerebral somite; prostomium.

Acrosternite: The narrow marginal flange anterior to the antecosta of a definitive sternal plate that includes the preceding primary intersegmental **Sclerotization:** characteristic of abdominal sterna of insects, but absent on thoracic sterna.

Acrostichal bristles: In Diptera, two rows of bristles on the middle of the dorsum of the mesothorax.

Acrotergite: The anterior precostal part of the tergal plate of a secondary segment; usually a narrow flange, but sometimes greatly enlarged, and frequently reduced or obliterated.

Acrotrophic egg tube: A type of egg tube in which the trophic cells remain in the apical (Telo trophic type).

Acrylonitrile: 1. Fumigant gas effective against nearly all pests. 2. Highly flammable and toxic. Acrylonitrile (66%) always must be mixed with carbon tetrachloride (34°/) before use. This mixture is a good spot fumigant that is effective against nearly all insects and is safe enough to apply to food processing areas however, it must be handled carefully. The person, applying it should wear a respiratory mask and should make sure that people, pets or wildlife are not exposed to the gas.

Activated Charcoal: Very finely ground, high quality charcoal which absorbs liquids and gases very easily.

Active ingredients: 1. Toxic component of a formulated pesticide. 2. The biologically active portion of a pesticide present in a formulation. 3. The chemical or chemicals in a product responsible for the desired effects, which are capable, in themselves, of preventing, destroying, repelling or mitigating insects, fungi, rodents, weeds, or other pests.

Actual Dosage: The amount of active ingredient (not formulated product) which is applied to an area or other target.

Aculea (pl: aculeae): A fixed hair of the cuticula.

Acute dermal toxicity: Poisoning from a single dose of a chemical absorbed through the skin.

Acute inhalation toxicity: How poisonous a single dose (or exposure) of a pesticide is when breathed into the lungs.

Acute oral toxicity: Poisoning from a single dose of a chemical taken by mouth.

Acute poisoning: Poisoning which occurs after a single dose (or exposure) of a pesticide.

Acute Toxicity: 1. How poisonous a pesticide to an animal or person after a single dose (or exposure). 2. Rapid poisoning produced by a single exposure or a few repeated exposures to excessive amounts of a toxic material.

Adductor muscle: Any muscle that draws an appendage to the body or brings parts into apposition.

Adfrontal areas: Two narrow oblique plates on head of lepidopterous larvae.

Adjuvant: 1. A spray additive to improve either physical or chemical properties. 2. A chemical or agent added to a pesticide mixture which helps the active ingredient do a better job. Examples: Wetting agent spreader, adhesive, emulsifying agent, penetrant. 3. Accessory substances; ingredients that improve the properties of pesticide formulations. They include Wetting agents, spreaders, emulsifiers, dispersing agents, etc.

Adrenergic blocking-agents: Agents that block the nerve fibers which liberate sympathin an adrenaline like substance.

Adsorption: The process by which materials are held or bound to the surface, in such a manner that the chemical is only slowly available. Clay and high organic soils tend to adsorb pesticides in many instances.

Adult: A full grown, sexually mature insect, nematode, or other animal. This stage is often the one to migrate and begin new colonies or infestations.

Adulticide: An insecticide which is toxic to the adult stage of insects.

Adultoid reproductive: In the higher termites, a supplementary reproductive that is a fully developed imago and is thus morphologically indistinguishable from the primary reproductive it replaces.

Adventitious vein: A secondary vein, neither accessory nor intercalary, usually the result of a number of cross veins lined up to form a continuous vein.

Adeeagal apodeme: An apodeme of the aedeagus.

Aedeagus: This is a copulatory organ of a male insect; intromittent organ, penis, or phallus.

Aerial: Flying forms.

Aerobic: Living or functioning in air or free oxygen. The opposite of anaerobic.

Aerobic respiration: The oxidative process of releasing energy within the cell.

Aeropyle: Specialized pores on the egg surface for oxygen uptake,

Aerosol: A dispersion of spray droplets of diameters in the range of 0.1-50 μm.

Aestivation: Dormancy during a hot or dry season.

Afferent nerve: A nerve that conducts from the periphery toward a nerve center; the axon of a sensory neutron.

Aflatoxins: Organic compounds (metabolites) produced by the fungus Aspergillus flaws which are highly toxic and carcinogenic to mammals.

Agroecosystem: It is a complex of organisms and the environmental factors found in an agricultural field, It differs from a natural ecosystem in that if is relatively artificial.

Air sac: A thin-walled dilation of the trachea. It lacks taenidia in its walls. It is capable of inflation and supposed to lessen specific gravity, and to aid in flight.

Alarm pheromone: A chemical substance exchanged among members of the same species that induces a state of alertness or alarm in the face of a common threat.

Alarm-defense system: Defensive behavior which also functions as an alarm signaling device within the colony. Examples include the use by certain ant species of chemical defensive secretions that double as alarm pheromones.

Alarm-recruitment system: A communication system that rallies nest mates to some particular place to aid in the defense of the colony. An example is the odour trail system of lower termites, which is used to recruit colony members to the vicinity of intruders and breaks in the nest wall.

Alary muscles: The dorsal transverse muscles of the body, attached medially on the heart, usually arranged in fan-shaped groups of fibers. (Wing muscles of the heart).

Alate: Possessing wings.

Alcohol: Although in ordinary conversation ethyl alcohol is generally referred to merely as 'alcohol' that term applied to a long series of hydroxy organic compounds beginning with the one-carbon compound methanol, or methyl alcohol. Both methyl alcohol and ethyl alcohol are common solvents, frequently used in formulating pesticidal mixtures.

Aldicarb (temik): Highly toxic. Temik is used chiefly as a soil treatment pesticide for sugar beet root maggots and as a soil-applied plant systemic insecticide for insect pests on potatoes and some greenhouse plants.

Alimentary Canal: The tube between the mouth and anus in which food movement, digestion, and absorption occurs

Alinotum: The wing-bearing plate of the dorsum of the mesothorax or metathorax of pterygote insects.

Allelo-chemical: Substance, emitted by an organism, that mediates inter specific interactions among organisms i.e. modifies the behavioural response in species other than the emitter. Depending on the type of the response induced, these chemicals have been classified into four groups viz. allomones, apneumones, kairomones and synomones.

Allethrin: 1. Low toxicity general purpose pesticide noted for its quick knock-down and insect flushing action. Allethrin is highly effective against some insects, but less effective against others. 2. Low toxicity. Allethrin is a synthetic (artificial) pyrethrin. Allethrin must be used selectively since it is highly effective against some insects but less effective against others. It is especially useful as a "quick knock down" insecticide for almost any type of insect or arachnid pest. 3. A synthetic insecticide related to the plant derived pyrethrins; relatively harmless to warm blooded animals, acute oral LD_{50} for rats 920 mg/kg.

Allomone: A chemical substance produced or acquired by an organism, which when it contacts an individual of another species in the natural context, evokes in the receiver a behavioural or physiological reaction adaptively favourable to the emitter.

Allopatric: Geographically isolated.

Alternative host: Host plants other than the main host on which an organism (bacterium, fungus, insect, virus etc.) can develop to complete its life cycle. Weeds and wild plants are often alternative hosts of certain pests and diseases.

Altosid (Methoprene): Low toxicity Altosid is a slow-release larvicide that is used by mosquito control districts to kill mosquito larvae in water before they mature to adults. Altosid can be used to kill mosquito and other water breeding insects around the house, particularly in small ponds, pools, ditches, etc.

Amaze: Highly toxic Amaze is used principally as a soil insecticide against the corn rootworm.

Ambulatory legs: Legs adapted for walking.

Ametabola: Insects, such as silver fish, which develop with slight or without metamorphosis. It is a primitive condition and the adult closely resembles the young stage.

Amnion cavity: Amniontic cavity, a space or cavity, in the developing egg, formed by the amniotic folds containing the germ band.

Amphibians: Animals of the class Vertebrata that are intermediate between fish and reptiles. They are cold blooded and have moist skin without scales, feathers or hair. Examples are frogs, toads, and salamanders.

Amphipneustic: An insect respiratory system (particularly in some dipterous larvae) in which only the first anterior and last posterior pairs of spiracles are functional.

Anaerobic: Living or functioning in the absence of air or free oxygen.

Anaerobic respiration: Releasing energy within the cell without oxygen.

Anal area of the wing: The posterior portion of the wing, usually including the anal veins.

Anal cell: A cell in the anal area of the wing; cell *I*A.

Anal crossing: Where vein A branches posteriorly from Cu+A.

Anal fan: A fan-like extension of the anal area in the hind wing of insects.

Anal gland: Ectodermal glands of the alimentary canal opening posteriorly or the anus.

Anal lobe: Basal part of wing behind the anal vein or posterior lobe of the wings of insects.

Anal loop. A group of cells in the hind wing of dragonflies, between Cu_2, *I*A and 2A, which may be rounded, elongate, or foot-shaped.

Anal margin. Margin of the wing near the anus of an insect.

Anal veins: All the veins between the cubitus and the jugal region, including, according to the Com stock-Needham system, the veins here called postcubitus and vannals.

Anamorphic development. Development by adding segments to the body.

Anaphylaxis. Hypersensitivity to a foreign protein or poison (such as a venom) marked by an intense systemic reaction.

Anapleurite: The dorsal supracoxal sclerotization of a generalized thoracic pleuron. (Eupleuron.)

Angstrom. One hundred millionth of a centimeter, a unit used in measuring the length of light waves.

Anionic. An ion having a negative charge is an anion. When the surface active portion of surfactant molecule possesses a negative charge it is termed and anionic surface active agent. Contrest with cationic.

Ant-lion. It is a predatory larva of a winged insect resembling a dragonfly. It digs a pit in sandy soil and lies with its large jaws partly exposed as it wants for passing insects particularly ants.

Antagonism. The phenomenon which results in a depression of compound activity, when two or more are in close, proximity or mixed together.

Anteclypeus. An anterior division of the clypeus.

Antecosta: The anterior sub marginal or marginal ridge on the inner surface of a tergal or sterna plate corresponding to the primary intersegmental fold, on which typically the longitudinal muscles are attached.

Antecostal suture: The external groove of the antecosta.

Antecoxal sclerite: A sclerite of the metasternum, just anterior to the hind.

Antenna (pl: Antennae): Feeler; a pair of flexible, many-jointed filaments articulated with head in procephalic region and bearing sense organs (tactile, auditory, olfactory). **Acuminate:** ending in a point. **Bipectinate:** fringed (pectinate) on both sides, (Lymantriddae). **Capitate:** ending in a capitulumor Knob (khapra beettes). **Clavate:** club-shaped, gradually clubbed (Curculionidae) **Elbowed (Geniculate):** bent at union of scape and pedicel (Chalcidoidea) **Filiform Thread :** like (Gryllidae). **Genniculate:** abruptly bent (Formicoidea). Knobbed swollen-tipped abruptly clubbed (Papilionidae). **Lamellate:** expanded in flattened plates (Melolonthidae) **Moniliform:** like a string of beads (termites-Isoptera) **Pectinate:** toothed like a, comb. (Lasicampidae and Sturniidae) **Plumose:** feather like (mosquito) Serrate: notched like, a saw (Elateridae) **Setaceous:** bristle-like (locust, grasshopper).

Anterior Notal Wing Process: The anterior lobe of the lateral margin of the alinotum supporting the neck of the first axillary (Vorderer Tergalhebel.)

Anti feedant: A chemical possessing the property of inhibiting the feeding of certain insect pests.

Anti-frothing agent: Material added to prevent frothing of the liquid in the spray tank.

Antibiosis: 1. An association between two or more organisms that is detrimental to one or more of them. 2. The resistance of a plant to Insect attack by having for example, a thick cuticle, hairy leaves, toxic sap, etc.

Antibiotics: Substances that are "against life." They are chemical substances produced by certain living cells such as bacteria, yeasts and molds. They are antagonistic or damaging to other Jiving cells such as disease producing bacteria multiplying. Penicillin is an example of an antibiotic that damages certain bacteria that cause disease in man.

Anticoagulant: A chemical used in a bait to destroy rodents. It destroys the walls of the small blood vessels, and keeps the blood from clotting; As a result the animal bleeds to death.

Antidote: A practical immediate treatment including first aid in case of poisoning a remedy used to counteract the effects of poison.

Anus: Posterior part; end opening of the alimentary canal.

Aphicide: An insecticide specially affective against sap sucking plant lice or aphids.

Aphid: A plant louse, also called greenfly, Aphids are small, soft skinned, often green plant-bugs. They suck sap mainly from young leaves and shoots, causing reduced growth and vigour, and leave a sweet, sticky excretion (honeydew) which is eagerly eaten by ants and which occludes the leaf and shoot pores. Often found in large numbers on wild and cultivated plants. Many are pests, e,g. Bean Aphid.

Apiary: 1. A place where honeybees are kept. Specifically, a group of hives. 2. Colonies of bees in hives and other bee-keeping equipment in a location and used for the purpose of producing honey, worker bees or queen bees.

Apical: At the end, tip, or outermost part.

Apiculture: The science and art of raising honey bee; bee-keeping; bee culture.

Apiology: The scientific study of bees, especially honey bees.

Apneustic: Without specific external breathing organs, either spiracles or gills; the tracheal system usually absent or rudimentary.

Apodeme: Any rigid process forming the insect endoskeleton; an invagination of the body wall.

Apodous: Without feet; footless.

Apophysis (pl: apophyses): A tubercular or elongate process of the body wall, either external or internal.

Appendage: Any part, piece or organ attached by a joint to the body or to any other main structure.

Appendiculate: Bearing appendages.

Applied control, artificial control: The measures adopted by human beings to control pests.

Applied entomology, economic entomology: The application of pure entomology to the control of insect pests.

Apposed image, apposition image: An image built up in eyes adapted solely for seeing by day, by apposed points of light falling side by side and not overlapping; mosaic image.

Apposition eye: In diurnal insects, an eye which absorbs oblique rays of light in the pigmented cells of the ommatidia.

Apposition image: In diurnal insects an image built up in eyes by apposed points of light falling side by side and not overlapping.

Apterous: Without wings; wingless.

Apterygota: A subclass of Insecta. It includes insects, such as springtails, bristle-tails and silver fish, which are primitively wingless and do not undergo, **Metamorphosis**. It does not include bugs, fleas and lice that are secondarily wingless owing to their parasitic habit. Because of many similarities, they are probably; descendent of crustaceans.

Arista: A specialized bristle or process on the antenna of certain Diptera; usually dorsal or terminal.

Arolium: A cushion like pad located between the tarsal claws and comprising part of the pretarsus.

Arota: The forepart of dorsal blood vessels of insects. In contrast to the hind part (heart), the arota is without valves. It discharges into the head region.

Arrhenotoky: A facultative type of parthenogenetic reproduction in which only male progeny are produced.

Articulation: 1. A movable point of contact between two sclerotic parts of the body. 2. A joint, as between two segments or structures. 3. Movable point or place where two segments or parts are joined.

Artificial infestation: Infestation of plants with pests by placing them directly on plants by hand or by releasing them into a cage, in contrast to natural infestation where a natural field population is depended on for testing a pesticide.

Asexual: Without sex; having the reproductive organs incompletely developed and producing eggs or young by cell-budding; parthogenetic.

Aspirator: A device with which insects may be picked up by suction.

Atomiser: Devise for breaking liquid stream into a very fine droplets by the high pressure of compressed air.

Atomization: Process of breaking a liquid into a fine spray.

Atrial orifice: The external opening of the spiracular atrium (Porta atrii.).

Atrium: The spiracular chamber formed by a secondary invagination of the body wall external to the primary tracheal orifice.

Atropine: A poisonous, crystalline alkaloid used in medicine; a specific antidote for poisoning by organic phosphate insecticides.

Atropine sulfate or atropine: An antidote used by doctors to treat people or animals poisoned by organic phosphate and carbamate pesticides in an attempt to save their lives.

Attapulgite: A magnesium silicate clay that is used as a carrier for the insecticide.

Attractants: 1. Materials that attract insects or other animals and cause them to eat or contact poison bait or sprays, and consequently cause their death. 2. Substance that attracts certain insects by some mechanism like, olfactory, gustatory, vision, 1actile etc. 3. Substances which elicit a positive directive response; chemicals having positive attraction for animals such as insects, usually in low concentration and at considerable distances. 4. Substances or devices capable of attracting insects or other pests to areas where they can be trapped or killed.

Auditory organs: Organs for hearing.

Auricle: A small lobe or earlike structure.

Autocidal control: The use of an insect species against itself, usually through some means of genetic modification, to suppress or eradicate its natural population.

Autolysis: The self destruction of biological cells after death as a result of the action of their own enzymes.

Autoparasitism (adelphoparasitism): A special type of hyper parasitism in which the female develops as a primary parasitoid, but the male is a secondary parasitoid through females of its own species.

Auxilary vein: The subsosta.

Average particle size. a. The arithmetic mean diameter; b. the diameter of particles of average surface; c. the diameter of particles of average volume; d. the diameter of particles of same specific surface as powder (the particulate diameter measured should be clearly stated).

Axilla (pl:, axillae): A triangular or rounded sclerite laterad of the scutellum, and usually just caudad of the base of the front wing.

Axillary cord: The thickened, corrugated posterior edge of the articular membrane of the wing base, continuous with the posterior marginal fold of the alinotum.

Axillary plate: The posterior sclerite of the wing base in Odonata, supporting the subcostal, radial, cubital, and vannal veins.

Axillary region: The region of the base containing the axillary sclerites.

Axillary sclerites: The sclerites of the axillary region in the wing-flexing insects, partly differentiated in Ephemerida, represented by the axillary plate in Odonata.

Axon: The process of a nerve cell that conducts impulses away from the cell body.

Azinphosethyl (ethyl guthion): A synthetic insecticide and acaricide; an organophosphate, O-diethyl S-4-oxo-1,2, 3-benzotriazin-3(4H)-ylmethyl phosphorodithioate; highly toxic to mammals; acute oral LD_{50} for rats 7 to 18 mg/kg; used on cotton and potatoes. Highly toxic Guthion is a general usage, broad spectrum insecticide that is effective against a wide variety of insects on fruits, vegetables, flowers, and ornamental plants.

Bacteria: Organisms (germs), some of which cause diseases in plants or animals. They are too small to be seen without a microscope.

Bait: 1. Components of food used for enticing the insects. 2. Foodstuff used for attracting pests; usually mixed with a poison to form a poison bait. 3. An edible material that is attractive to the pest, which normally contains a pesticide unless used as a prebait.

Bait shyness: 1. The tendency for rodents, birds, or other pests to avoid a **Poisoned bait**. 2. The discarding of a poison bait by the animals after the initial mortality of their fellows.

Band application: 1. An application to a continuous restricted area such as in or along a crop row rather than over the entire field area. 2. Treatment of a band of soil in row crops, usually covering plant rows, with either sprays or granules.

Barrier application: The use of pesticide or another agent to stop pests from entering a container, area, field or building.

Basal apodeme: An apodeme of the phallobase.

Basal treatment. A treatment applied to the stems or trunks of plants at and just above the ground line.

Basalare: The episternal epipleurite (sometimes double) giving insertion to the anterior pleural muscles of the wing; often represented by an undetached or partially detached lobe of the episternum before the pleural wing process (Episternal-gelenkstuck, preparapteron).

Basement membrane: Thin noncellular membrane forming the inner lining of the hypodermis of the body wall.

Basicosta: The proximal sub-marginal ridge of the inner wall of a leg segment.

Basicostal suture: The external groove of a leg segment forming the basicosta.

Basicoxite: The usually narrow basal rim of the coxa proximal to the basicostal suture and its internal ridge. (Coxomarginale).

Basisternum: That part of the sternum anterior to the sternacostal suture.

Basitarsus: The proximal or basal segment of the tarsus.

Bee: A furry insect of the order Hymenoptera. There are solitary and social bees. The latter live in colonies. Each colony has one Queen bee together with large numbers of female 'workers' and fewer 'drones'. The most highly socialized is the Honey bee (*Apis mellifera)* kept in hives by beekeepers for commercial. Honey production, Bumblebees (Bombus spp.) have much samller colonies. A growing practice is the hiring and placing of Beehives in Orchards to promote pollination and fruit setting.

Bee farming: The keeping of bees for honey and beeswax they produce, and for their pollinating of plants.

Beehive: A box in which bees are kept to produce Honey. Several types of single and double hives are in common use, the former favoured for commercial production. Hives contain two separate chambers, a lower brood chamber in which the Queen lays and rears her Brood, above which is the honey chamber or supers containing frames in which the bees store surplus honey which periodically removed.

Beeswax: A mixture or organic compounds secreted by special on the last four segments on the ventral side of the worker bee's abdomen; used by the bees-for building comb. Byproduct of bee-keeping industry; produced from honeycombs, mostly from wild bee combs; yellowish to grey brown solid, with agreeable honey like odour; brittle when cold, turns plastic by heat of the hand; insoluble in water and soluble in water and soluble either, chloroform and fixed and volatile oils,; used for cosmetics candles and a number of industrial products.

Behavioural resistance: A mechanism in insects of resistance to insecticides which involves a behavioural change whereby contact with the insecticides is avoided.

Bendiocarb: Highly toxic. This is a wide spectrum insecticide that is especially useful against soil insects. It provides good residual action.

Bentonite: A clay composed mainly of silica and aluminum silicate; mined in Mississippi, Wyoming and elsewhere; used as dust diluents and for lining ponds to hold water.

BHC (Common name): A synthetic insecticide, a chlorinated hydrocarbon, 1, 2, 3, 4, 5, 6, -hexachlorocyclohexane, mixed isomers and a specified percentage of gamma; slightly more toxic to mammals than DDT, acute oral LD_{50} for rats about 200mg/ kg; phytotoxicity more toxic that DDT, interferes with germination, suppresses growth and reduces yields except at low concentration certain crop plants as potato absorb crude SHC with consequent tainting of tubers; product of several companies.

Biforous spiracles: Spiracles of coleopterous larvae having two pouches of the atrium originally supposed to open separately to the exterior.

Bilateral symmetry: Symmetry in which the body can be divided by a median plane into mirror images of one another.

Binapacryl (morocide): A synthetic acaricide; a nitrophenol derivative, 2-sec-butyl- 4,6,- dinitrophenyl 3-methyl-2-butenoate; moderate mammalian toxicity; acute oral LD_{50} for rats 136 to 225 mg/kg; controls mites on tree fruit and huts; product of Niagara.

Binomial: Pertaining to two names, the zoological system of nomenclature consisting of a generic and a specific name.

Binomial nomenclature: The method of naming plants and animals introduced by Linnaeus, a Swedish naturalist. Every plant or animal has two Latin names, a generic name designating its genus and a specific name indicating the species.

Bioaccumulation of pesticides: Accumulation of the toxicant inside dynamic ecological system depending on (i) physicochemical characteristics of Insecticide; (ii) concentration vs dilution or degradation in the system; (iii) rate of food consumption and body size and (iv) dynamics of accumulation i.e balance of uptake and elimination.

Bioassay: Estimation of amount of toxicant in a sample by measurement of its effect on test organisms: usually estimation of amount necessary in a standard type of application to produce 50% kills.

Biocide: 1. A chemical which has a wide range of toxic properties, usually to members of both the plant and animal kingdoms. 2. General toxicant used for killing.

Bioconcentration: 1. The process of a pesticide becoming concentrated in plants or animals. When a chemical increases in concentration at each succeeding link in the food chain. 2. Capable of getting broken down in the biological system.

Biological control: 1. Biological pest suppression in its narrow, classical sence usually restricted to the introduction, by man of parasitoids, predators, and or pathogenic microorganisms to suppress populations of plant or animal pests. 2. The control of pests by employing predators, or- disease; the natural enemies are encouraged and disseminated by man.

Biological diversity: The number of different kinds of organisms per unit area of volume the richness of species in a given area.

Biological magnification: When a small organism feeds on several Micro-organisms containing residues of pesticide; the pesticide may get deposited in its body. A predatory animal which feeds on several such small organisms ingests a larger quantity of the pesticide. In this way, the larger beings at the top of the food chain consume quantity of the pesticide much larger than that was present in the microorganisms.

Biological oxygen demand: The quantity of oxygen used by microorganisms in the biochemical oxidation of organic matter and oxidisable inorganic matter by aerobic biological action.

Biomagnification, pesticides: In certain aquatic animals and plants the residue levels of pesticides are much higher than in other plants or water itself e.g. in snails and phytoplankton. It is because they accumulate the lipid soluble pesticides both from water and food chain. This process is called biomagnification of pesticide/residues.

Biomass: The total dry weight or volume of all the organisms in a given area.

Bionomics: The study of the habits, breeding and adaptations of living forms.

Biotic: Relating to life.

Biotic potential: The maximum possible rate of increase (multiplication) of an organism in the absence of any limiting factor or strress. Also called Absolute Reproductive Potential.

Biotype: A biological strain of an organism, morphologically indistinguishable from other members of its species but exhibiting distinctive physiological characteristics; particularly in regard to its ability to successfully utilize pest - resistant host organisms or to act as an effective beneficial species.

Bipectinate: Antennae having two rows of comb like processes or teeth; bushy or feathery antennae.

Bisexual: Having two sexes distinct and separate; with males and females.

Blanket application: The application of spray or dust over an entire area rather than on rows of beds.

Blastocoel (Be): A fluid-filled cavity within a blastula.

Blastoderm: The cells produced through cleavage of the zygote.

Blastogenesis: The origin of different caste traits from variation in either the ovarian environment of the egg or the non genetic contents of the egg (opposed to genetic control of caste and trophogenesis).

Blastokinesis: Movement or migration of the embryo within the egg.

Blastomeres: The cleavage cells, or cells produced by the division of the egg or its nucleus that form the blastoderm.

Blastopore: The mouth of the gastrulation cavity.

Blastula: The early stage of the embryo in which the only cell layer is the blastoderm.

Boom: A bar held parallel to the ground during operation, on which are mounted, at regular spacing, the nozzles of a crop sprayer or duster. The boom is often hollow, serving also as a supply tube for the spray liquid.

Bordeaux mixture: Primarily a fungicide but also a repellent to many insects; a popular formula consists of copper sulfate 6 lb, hydrated lime 10 lb, and water 100 gal; ingestion of large quantities may cause fatal gastroenteritis in mammals; toxic to some plants, particularly at low temperatures.

Brachypterous: Insects with short wings which do not cover their abdomen fully.

Broad spectrum insecticide: Non-selective, having about the same toxicity to most insects.

Bromophos. **Moderately toxic:** This is a wide-spectrum insecticide that is especially useful against sucking and biting insects such as those on flowers, plants, and shrubs.

Brood cell: A special chamber or pocket built to house immature stages.

Brood comb: A wooden frame enclosing a wax sheet from which bees build the cells of the honeycomb, placed in the brood chamber of a hive.

Buccal cavity: The first part of the stomodaeum, lying just within the mouth; its dilator muscles arising on the clypeus, and inserted before the frontal ganglion and its connectives.

Bucket pump: The simplest hydraulic sprayer is the bucket pump. It is a plunger pump adapted by a clamp or otherwise for use in an open pail and delivering the liquid through a spray nozzle at the end of a hose.

Bucket sprayer: Sprayer designed to be used with a bucket containing spray liquid; provided with single or double - acting pump; used for spraying small areas in garden or nursery beds; laborious for use over field crops.

Build up: Accumulation of a pesticide in soil, animals, or in the food chain.

Bursa copulatrix: A copulatory pouch of the female, usually the genital chamber or a part of the latter.

Caecum (pl:, caeca): 1. A tube like blind structure which arises from the anterior end of the mid-gut in certain insects. 2. A blind sac, one of a group of appendages opening into the anterior region of the ventriculus.

Campaniform sensillum: A sense organ consisting of a dome shaped cuticular area into which the sensory cell process is inserted.

Campodeiform larva: A larva shaped like the thysanuran Compodea, that is elongate, flattened, with well-developed legs and antennae, and usually with filaments on end of abdomen, and usually active.

Cannibalistic: Feeding on other individuals of the same species.

Capitate: With a head, a type of antenna terminating in a knob-like process.

Capitulum: Head-like structure of ticks which bears the feeding organs.

Capsule: The "granular" inclusion body which is characteristic of a granulosis viral infection; the proteinaceous covering of a granulosis virus rod produced in the infected tissue cell.

Carabiform larva: A larva shaped like the larva of a carabid beetle, that is elongate, flattened, and with well-developed legs; filaments lacking on end of abdomen.

Carbamates: 1. A group of alkaloid insecticides which are selective and to some extent systemic in their action, and are non-persistent in the environment, e.g. carbaryl methiocarb. 2. A group of chemicals which are salts or esters of carbonic acid. Includes insecticides, herbicides, and fungicides.

Carcinogenic: The term used to describe the cancer producing property of a substance or agent.

Cardiac sinus: The channel of the haemocoel dorsal to the yolker alimentary canal.

Cardiac valve: A value at the junction of the proventriculus and ventriculus.

Cardioblasts: The dorsal strands of mesodermal cells that form the dorsal blood vessel.

Carda (pl: cardines): The proximal piece of the maxilla.

Carnivore: An animal (or, rarely, a plant) which eats animal flesh. Adj. carnivorous.

Carpopodite: The fifth segment of a generalized appendage (Tibia).

Carrier: 1. An inert material that serves as a diluent or vehicle for the active ingredient or toxicant. 2. The liquid or solid material added to a chemical compound to facilitate its field application. An inert material which when used with a toxic compound improves the physical dispersion of the toxicant.

Caste: 1. Broadly defined, as in ergonomic theory, any set of individuals of a particular morphological type, or age group, or both, that performs specialized labour in the colony. More narrowly defined, any set of individuals in a given colony that are both morphologically distinct and specialized in behaviour. 2. A form or type of adult in a social insect such as among termites and ants.

Catalyst: A substance that speeds up the rate of a chemical reaction but is not itself used up in the reaction.

Caterpillar: 1. Worm, like larva of a butterfly or certain other insects 2. lepidopterous larva well developed head, 3 thoracic and 10 abdominal segments, a pair of 5 jointed legs on each thoracic segment and 5 pairs of legs (prolegs) on III, IV, V, VI and X abdominal segments. In geometridea, prolegs are generally present only on VI and X segment (looper) while in Plusiinae, 1st or 1st and 2nd pairs are aborted (semilooper). Larva of Microteryx spp. (Micropterygidae) has 8 pairs of prolegs while that of Phyllocnistis spp; (Gracilariidae) is apodous.

Cationic: An ion having a positive charge is a cation. When the surface active portion of a surfactant molecule possesses a positive charge it is termed a cationic surfactant.

Caudal: Pertaining to the posterior region or to the anal end of the insect body.

Cement layer: A thin layer on the surface of insect cuticles formed by the hardened secretion of the dermal glands.

Cenchrus (pl: cenchri): A pale coloured membranous lobe or area on each side of the metanotum of Chalastogastra (Hymenoptera).

Cephaled: Toward of front of an animal.

Cephalic: Pertaining to the head or attached to the head.

Cephalic lobes: The head lobes of the embryo, comprising the region of the prostomium and usually that of the tritocerebarl somite

Cephalothorax: A body region consisting of head and thoracic segments, as in spider.

Cercus (pb cerci): An appendage usually paired of the tenth abdominal segment, usually segmented and slender.

Cerebral ganglion: Supra oesophageal ganglion. The brain of an insect. It lies between the supporting apodemes of the tentorium, just above the oesophagus.

Cervical: Pertaining to the neck or cervix.

Cervical sclerites, cervicalia: The sclerites of the neck; particularly one or two pairs of lateral neck plates (Kehlplatten) joining the head to the prothoracic episterna.

Cervix: The neck; including probably the posterior non-sclerotized part of the labial somite and the anterior part of the prothorax,

Chaetotaxy: The arrangement and nomenclature of the setae or bristles on any part of the exoskeleton, especially in the Diptera; the study of such an arrangement.

Chelicerae: The first pair of appendages of adult Chelicerata, innervated from the tritocerebral ganglia of the brain; equivalent to the second antennae of Crustacea.

Chemoreceptor: A sense organ sensitive to chemical properties of matter (a taste receptor or an odour receptor).

Chemosterilant: Chemical used for making an insect sterile without killing it.

Chemotherapy: The treatment of a desired plant or animal with chemicals to destroy or control a pathogen without seriously harming the plant.

Chironomid: An insect of the order Diptera whose young ones; blood red and worm-like, are invariably found at the bottom of ponds living in tubes.

Chitin: 1. A nitrogenous polysac charide occurring in the cuticle of arthropods and certain other invertebrates. Probably occurs naturally only in chemical combination with protein. 2. A nitrogenous polysaccharide found in the exoskeleton of insects and in the cell walls of many fungi. It provides mechanical strength and resistance to chemicals.

Chloropicrin: 1. Insecticidal fumigant for treatment of stored grain, and cereal products and for soil treatment to control insects, nematodes, weeds, fungi; empirical formula CCl_3, $N0_2$; 0.05 oz/ 1000 cu ft lethal in 10min to mammals; very toxic to plants when injected into soil. 2. Fairly toxic. "Tear gas" should be handled with caution, It is a good spot fumigant against a variety of pests, and is used as a warning gas for other more toxic, but less detectable fumigants, 3. Also known as tear gas; clear to yellow, volatile soil fumigant; a nematocide, rodenticide and herbicide. Poisonous to humans.

Chlorosis: In plants, yellowness of normally green, tissues, due to partial failure of chlorophyll to develop or to removal of chlorophyll

Chlorpyrifos (Dursban; Lorsban): 1. Moderately toxic Dursban is especially effective against cockroaches and many other insects as well. (Although it acts somewhat slowly) Dursban provides good residual action, particularly on smooth and non-porous surfaces such as stainless steel, linoleum, tile, etc. 2. Though somewhat toxic, Dursban is especially effective against cockroaches and many other insects as well, both indoors and outdoors. Dursban acts somewhat slowly. Provides good residual action, particularly on smooth and non-porous surfaces such as stainless steel, linoleum, tile, etc.

Chromosomes: At cell division the dark-staining, rod-shaped structures which contain the hereditary units called genes.

Chronic effect: A slow and long continued effect.

Chronic poisoning: Resulting from long periods of exposure to a chemical.

Chronic toxicity: Ability to cause injury or death from prolonged exposure.

Cibarium: The food pocket of the extraoral or preoral mouth cavity between the base of the hypopharynx and the under surface of the clypeus.

Circadian rhythm: Physiological rhythm with an approximate 24 hour cycle.

Circumoesophageal connectives: The connectives between the brain and the ventral nerve cord embracing the stomodaeum. Primitively the connectives from the archicercbrum to the first ventral ganglia; in insects the connectives between the tritocerebral ganglia and the mandibular ganglia.

Clasper: One of a pair of appendages at the tail end of a caterpillar by which it clings to leaves, twigs and other surfaces.

Classification: 1. The process of assigning pesticides into groups according to common characteristics. 2. The systematic arrangement of insects (or other animals or plants) in series showing their relation or agreement in structure, life habits or other characters forming the basis of classification.

Clavate: Club-like; thickening gradually toward the tip.

Clavus: The sharply pointed anal area of the hemelytra of the Heteroptera.

Cleavage: In embryological development, the division of the originally single-celled egg into a number of cells called blastomers.

Cleptobiosis: The relation in which one species robs the food stores or scavenges in the refuse piles of another species, but does not nest in close association with it.

Cleptoparasitism: 1. The parasitic relation in which a female seeks out the prey or stored food of another female, usually belonging to a different species, and appropriates it for the rearing of her own offspring. 2. A type of parasitism in which the adult parasitoid preferentially appropriates for its own progeny the previously paralyzed and parasitized host of another parasitoid (Arthur, 1961).

Clone: Offspring produced by asexual reproduction of single animal.

Clypeus: That part of the head in front of the frons and to which the labrum is attached.

Co-distillation: In the process of distillation, vapours are driven off by heat and then condensed. In co-distillation, an additional material is carried off with the vapour.

Coagulate: To change from a fluid to two or more species, none of which a jelly; to form a clot.

Coarctate pupa: A pupa enclosed of which is benefited inside a hardened shell formed by the previous larval skin.

Cochineal: A red dye that is ex- two ganglia of a segment or the tracted from the bodies of scale lateral centers within a median insects (Dactylopius coccus). About 500g of the dye is produced from 70,000 insects, living in a specific region.

Cockle: A small gall on wheat formed due to the attack from a nematode, *Auguina tritici.*

Cocoon: 1. The silky protective case of many larval forms before they become pupae; silky or other covering formed by many animals for their eggs. 2. A protective silky sheath spun by many insect larvae (e.g. the Caterpillars and moths) in which they pupate See PUPA). Also a similar sheath or capsule in which spiders and earthworms lay their eggs. 3. Protective case formed by larvae of various insects species within which they pass their pupal stage.

Coelome: A body cavity formed of the coelomic sacs only.

Coevolution: The evolution of a specific relationship between two or more distinctly different species.

Colleterial glands: Accessory glands of the female secreting an adhesive substance used to the eggs to a support.

Collophore: 1. A ventral tube on the first abdominal segment of Collem- bola. 2. A tube like structure located on the ventral side of the first abdominal segment of springtails.

Colon: The posterior part of the anterior intestine, between the ileum and the rectum. (large intestine, Dickdarm.)

Commensalism: A living together of two or more species, none of which is injured thereby, and at least one of which is benefitted.

Commissure: A transverse tract of nerve fibers connecting the ganglia of a segment or the lateral centers within a median ganglion.

Community: All plants and animals living in a specific region.

Compatability: 1. The ability to mix different pesticides without physical or chemical interactions which would lead to reduction in biological efficiency or increase in phytotoxicity. 2. The ability of a pesticide to mix with other pesticides without any undesirable effects, such as reduction in the pesticidal activity, increase in the plant injury, promotion of blockage of the spray nozzles, and excessive run-off from the treated surfaces. 3. When two materials can be mixed together with neither affecting the action of the other.

Complete metamorphosis: Holometabolic development; immature (larvae) do not closely resemble the adults and a pupal stage is present.

Compound eye: An eye composed of many individual elements or ommatidia, each of which is represented externally by a facet.

Condyle: A process which articulates the mandible to the head.

Control schedule. Time table for the application of control measures irrespective of the presence or absence of the pest (S).

Copulation: The union of the sexes for reproduction.

Corium: The elongate basal part of the front wing of Hemiptera.

Cornea: The outer surface of the compound eye as a whole, and of each individual facet; cuticular lens.

Corneagen layer: Transparent epidermal cells beneath the cuticular lenses (cornea) in the compound eyes.

Cornicle: Honey - tube (erroneous); a pair of tubes present on 5th abdominal segment of aphids (Aphididae - wanting in phylloxerinae), secretary channels, excreting waxy fluid (not honey - dew), for self protection against predators.

Coronal suture: The stem of the Y -shaped suture on the vertex and extending forward.

Corpora allata: Paired endocrine organs located just behind the brain; the Source of juvenile hormone.

Cosmopolitan: An species occurring very widely throughout the major regions of the world.

Costa: A longitudinal vein extending along the anterior margin of the wing.

Costal margin: The anterior margin of a wing.

Cover crop: A subsidiary, fast growing, and usually annual crop that is grown with a more slowly growing biennial or perennial crop.

Coverage: The amount of spread of a pesticide over a surface.

Coxa: The distal part of the coxopodite serving as the functional basal segment of the leg when separated from the proximal part (subcoxa).

Coxopleurite: The sclerite of a generalized thoracic pleuron adjacent to the dorsal margin of the coxa, bearing the dorsal coxal articulation; its anterior part becomes the definitive. (Eutrochantin, T rochantinopleura.)

Coxopodite: The primary basal segment of an appendage, representing the primitive limb basis.

Cramps: An involuntary and painful contraction of a voluntary muscle or group of muscles.

Crochets (pronounced croshays): Hooked spines at tip of the prologs of lepidopterous larvae.

Crop: A thin walled dilation of oesophagus in certain insects for storing food for sometime before passing it on.

Cross resistant: When a pest population which has become resistant to one pesticide also becomes resistant to other chemically related pesticides.

Cross veins: Short veins between the lengthwise veins and their branches, numerous in net veined wings, in others generally few and located in definite positions.

Cryptic colouration: Colouration of organism that blends with background.

Crystalline cone, crystalline body, vitreous body: A transparent refractive subcorneal part of an oval or conical body formed of cells or of cell products.

Cubitus: A longitudinal vein of an insect wing behind the media, usually two branched.

Cultural control: 1. Control measures including modifications of the planting, growing, cultivation, and harvesting of crops aimed at prevention of pest damage rather than destruction of an existing infestation. 2. The control of pests by suitably altering or modifying the usual agricultural operations. It involves the use of ordinary farm practices and farm machinery.

Cumulative effect: The result of some poisons which build up or are stored in the body so that small amounts eaten or contacted over a period of time can sicken or kill an animal or person. Examples of chemicals which accumulate are: anticoagulant rodenticides, mercury compounds, thallium sulfate.

Curative: Having the power to heal or cure.

Cursorial: Insects having legs adapted for running.

Cuticula: 1. Outer covering of the body wall of an insect composed of a non-cellular layer of chitin. 2. The outer protective envelope of any living organism. In the case of mammals or arthropods the alternative names "skin" or "integument" are often used, but "cuticle" invariably in respect of green plants. Plant cuticle is a macromolecular structure of predominantly paraffinic composition over-lying cellulose and pectin. 3. The outer non-cellular layer of the body wall which is secreted by the epidermis. Its outermost sub-layer (epicuticle) is a thinner and multilayered structure, which is impervious to water owing to the presence of lipoids. The middle sub-layer (exocuticle) is pigmented and contains proteins, chitin and hardening-substance sclerotin which gives rigidity. The inner most sub-layer (endocuticle) is unpigmented, and contains proteins and chitin. The sub-layer gives elasticity to the body wall. 4. The outer non-cellular layers of the insect integument secreted by the epidermis.

Cyclodiene insecticides: They are highly chlorinated cyclic hydrocarbons with endomethylene bridged structures. Such insecticides, e.g. chlordane, heptachlor, aldrin, dieldrin and endrin are prepared by the Diels - Alder reaction that was discovered by two chemists named Diels and Alder.

Cyhexatin (Plictran): Moderately toxic. Plictran is a relatively new pesticide that is selectively useful against mites. Plictran also acts as a fungicide that is effective against some molds, mildew, and other household fungi.

Cynosis: A dark bluish or purplish colouration of the skin and mucous membrane due to deficient oxygenation of the blood in the lungs or to an abnormally great reduction of the blood in its passage through the capillaries.

Cystocytes: The cells enclosing the germ cells in a gonadial tube; the follicle cells of the ovary, the cells or the testis.

Cytoplasm: The fluid or gelatinous content of living cells as distinct from the reproductive nucleus.

Damage threshold: Pest population above which crop losses are visible.

Deflocculating agent: A substance that prevents rapid precipitation (settling out) of solids in the liquid in the spray tank.

Degradability: The ability of a chemical to decompose or break down into less complex compounds or elements.

Degree of freedom: 1. Term used to denote the least number of independent variables defining the state of a system (e.g. the temperature and pressure in case of a gas) which must be given definite values before this state is completely determined. 2. The number of independent ways in which a molecule may possess translational, vibrational or rotational energies.

Delirium: A condition of extreme mental excitement, marked by a rapid succession of confused and unconnected ideas.

Dendrons or dendrites: Finely ramifying branches given off from a nerve cell.

Density dependent: Refers to mortality factors or processes in the environment which destroy an increasing percentage of the subject population as the numerical population density increases, and Vice Versa.

Density dependent factor: In ecology a factor which changes in intensity with changes in population density (e.g., lethal effects of a factor intensify as population numbers increase).

Density independent: Refers to mortality factors or processes in the environment which destroy a relatively constant percentage of the subject population regardless of changes in its density.

Density independent factor: A factor regulating activity or population of living organisms but whose influence is independent of the population density of that particular organism (concerned organism).

Density population: Number of individuals of a species that habitate or live in per unit area (generally per square kilometer) of any geographical region.

Denticulate: Bearing a row of very small teeth.

Deposit: Amount and pattern of active ingredient deposited per unit area of plant surface.

Deposit: Amount and pattern of spray or dust deposited per unit area of plant surface.

Dermal: Through the skin.

Dermal toxicity: 1. Toxicity of a material as tested on the skin. 2. Dermal toxicity is the passage of pesticides into the body through the unbroken skin. Most pesticides pass through unbroken skin to some extent. Many are not adsorbed readily unless in certain solvents. However, many including most organophosphates, will pass through unbroken skin as technical material and in all formulations. Dermal exposure results from spillage onto skin and clothing (including gloves and other protective equipment), from drift or from damaged or improperly maintained equipment. It is important to be aware of hazards resulting from the dermal toxicity of the materials you use. This is the greatest hazard to people handling pesticides.

Detergent: A chemical (not soap) having the ability to remove soil and grime. Detergents can be used as surfactants in some pesticide sprays.

Deterrent: A chemical substance that deters feeding or oviposition of an insect.

Detoxification: Ridding a toxicant of its toxicity by chemical or physical means; metabolism (mostly catabolism or degradation) or xenobiotics in an organism or plant into ultimate harmless products.

Deuterotoky: A type of parthenogenetic reproduction in which the progeny of unmated females may consist of both males and females.

Deutocerebrum: The part of the arthropod brain containing the first antennal nerve centers.

Deutonymph. The third instar stage in the development of mites and ticks of protonymph.

Diapause: 1. It delays growth and development which is not immediately referable to the adverse environmental conditions. It is, thus, not easily altered by change to a more favourable environment. It is an adaptive phenomenon or inborn rhythm which usually facilitates survival during unfavourable periods. 2. A state of an animal, such as an insect, in which a reduction of growth processes or maturation occurs which is not necessarily caused by immediate environmental influence, does not depend for its continuance on unsuitable conditions, and is not easily or quickly altered by change to a more favourable environment; frequently related to change in photoperiod. Once the state of diapause comes to an end, normal growth and development are resumed. 3. A temporary stoppage of development of an insect embryo or larva.

Diaphragm: A horizontal membranous partition of the body cavity.

Dicondylic joint: A joint with two points of articulation between the adjacent leg segments.

Digestion: Enzymatic hydrolysis of food into absorbable size molecules.

Dilator muscles: Muscles extending from the body wall to the alimentary canal; called also suspensory muscles.

Dimorphic: Occurring in two distinct forms.

Dimorphism: 1. In caste systems, the existence in the same colony of two different forms, including two size classes, not connected by intermediates. 2. A difference in size, form, or colour, between individuals of the same species, characterizing two distinct types.

Diphasic allometry. Polymorphism in which the allometric regression line, when plotted on a double logarithmic scale, "breaks" and consists of two segments of different slopes whose ends meet at an intermediate point.

Direct pest: An organism which causes immediate and direct damage to a marketable item, such as a fruit, even at low population density.

Discal cross vein: A cross vein behind the discal cell.

Discoidal cell: A ceil near the middle of the front wing.

Discoidal vein: The vein forming a continuation of the median vein beyond the end of the basal vein, and extending along the posterior margin of the first discoidal cell.

Dispersal: Movement of individuals out of a population (emigration) or into a population (immigration).

Dispersing agent: 1. Spray additive which enables formation of a stable suspension of solid particles in the liquid media. 2. A material that reduces the cohesive attraction between like particles. Dispersing and suspending agents are added during the preparation of wettable powders to facilitate wetting and suspension of the active ingredient. 3. Substance added in water dispersible insecticide powders and concentrate pastes to get uniform concentration of spray all over treated area. Common example: protective colloids such as sodium carboxy and methyl cell uloses.

Disruption of communication (confusion technique): The use of pheromones, pheromone analogs, or pheromone masking agents to saturate the atmospheric environment of a pest and thus block the use of some important signal (usually reproductive) for successful perpetuation of the species.

Dissipate: To get rid of by spreading out. Gases and vapours dissipate through the air.

Distal process: The peripheral branch or one of several distal branches of a sensory nerve cell.

Diurnal: Active during daytime hours.

Diverticula (pl:, diverticulae): Caecal tubes or pouches; branches from alimentary canal; extensions or evaginations.

Dockage: Foreign material in harvested grain such as weed seeds, chaff, and dust.

Dormancy: 1. A state of quiescence or inactivity. 2. Alive but not growing; quiescent; with suspended animation or development.

Dormant spray: A spray applied during the dormant season when the plant growth is inactive.

Dorsal: Top or uppermost; pertaining to the back or upper.

Dorsal blastoderm (DBl): The extraembryonic part of the blastoderm. (Serosa).

Dorsal diaphragm (DDph): The membranous and muscular sheets of tissue extending from the dorsal blood vessel to the laterodorsal parts of the body wall, separating the dorsal sinus from the perivisceral sinus.

Dorsal ocellus: The simple eye in adult insects and in nymphs and naiads.

Dorsal organ: In the embryo of an insect, the amnioserosal sac arising from the amnion or a mass of cells on the dorsal side produced by the invaginated serosa.

Dorsal shield scutum: Sclerotized plate covering all or most of the dorsal surface in males and the anterior portion in females, nymphs and larvae of hard-backed ticks.

Dorsal sinus (DS): The space of the definitive body cavity above the dorsal diaphragm and the heart. (Pericardial sinus.)

Dorsal trachea: The dorsal segmental trachea originating at a spiracle.

Dorsal tracheal trunk (DTra): A longitudinal dorsal trunk uniting the series of dorsal tracheae.

Dorsal vessel (DV): The dorsal blood vessel, consisting of the pulsatile heart and the non-pulsating aorta.

Dorso-pleural line: The line of separation between the dorsum and the pleural region of the body, often marked by a fold or groove.

Dorsum: The entire back of an animal above the pleural regions; or specifically, when qualified by the designation of a segment, the back region of a segment.

Dosage: A dose is a measured quantity, as of medicine, taken at one time or in one period of time.

Drift: 1. Movement of spray or dust material by wind or air currents outside the intended area, usually as fine droplets, during or shortly after application. 2. Spray or dust carried by natural air currents beyond the target area.

Drone: 1. The male bee, hatched from an unfertilized egg; larger than the worker bee; does not gather honey; has no sting. 2. A male social bee, especially a male honeybee or bumblebee.

Droplet density: Number of droplets required per unit area (generally per square centimeter) for effective coverage of target in a spray operation. This is of the order of 10-20 droplets in high volume or normal sprays, 20-30 droplets in herbicide sprays, 56-70 droplets in ULV and 30-50 droplets per square centimeter in low volume sprays.

Duct: A tube for carrying secretion from a gland to the point of discharge.

Ductless glands (endocrine glands): The glands, not joined by tubular organs, that secrete hormones in the body which are transmitted throughout the body via the blood stream.

Dunnage: Materials, such as wooden planks, loose wood of any kind and wheat straw, which insulate the stored material against the seepage of moisture from the sides and floor.

Dust: 1. Fine, dry particles of earth or other matter which can be wafted by wind. 2. A pesticide applied in dry state; application of such dust to plants, animals, fowls. 3. A pesticide formulation in dry, finely divided form, containing a low concentration of an active ingredient to be used without further preparation or dilution.

Dust formulation: Mixing of an insecticide when it is to be sold as a dust with a spedal quality talcum or some other ingredient in a known proportion to from a fine powder. Amount of insecticide in such a mixture generally five to ten percent.

Duster: A mechanical device, consisting of a blower, dust reservoir, and a nozzle, used in the application of insecticidal or fungicidal dust so that the dust is emitted in a blast of air. Mainly hand-driven; also power-driven.

Dusting: The application of insecticide or fungicide to crops in a dry, powdery state. It is less efficient than spraying, usually requiring about twice the amount of chemical per unit area.

Dyar's Rule: An empirical law based on the observation that the head capsule and certain other parts of caterpillars grow in geometric proression, increasing in size by a constant ratio at each moult. Thus if the lograthim of the size of the head is plotted against the number of instar, the resulting graph will be a straight line. Although not always accurate, this rule has sometimes been of use in deducting facts about the life history of an insect in cases where other is lacking, especially indicating when one instar has been overlooked.

Ecdysial suture: Weakened area of integument that splits during molting to permit the old exoskeleton to be shed.

Ecdysis (pl: ecdyses): 1. Molting; the process of shedding the skin. 2. The molting (shedding of the skin) of larval arthropods from one stage of development to another the final moult leading to the formation of the puparium or chrysalis.

Ecdysone: Insect molting hormone; a chemical secreted by the prothoracic glands after each molt or at the termination of diapauses, which stimulates growth and development of various somatic tissues and other physiological and morphological changes necessary to the continued repetition of the molting cycle.

Eclosion: 1. Process of an individual's emerging or hatching from an egg. 2. Emergence of the adult (imago) from the pupa; less commonly, the hatching of an egg.

Ecological efficiency: Per cent of energy in biomass produced by a trophic level that is incorporated into biomass by the next higher trophic level.

Ecological homologues: 1. Two allopatric species of organisms with identical or very similar ecological niche requirements, especially in regard to some environmental provision present in only a limited supply. 2. Species that have the same fundamental ecological requirements and therefore cannot occupy the same area indefinitely without one species displacing the other.

Ecological niche: 1. The place an organism occupies in its biotic relationships and physical environment as determined by its particular structural adaptations, physiological adjustment and developed behavioral patterns. 2. The way of life of an organism including food, habitat requirements, and so forth.

Ecology: The study of organisms in relation to their environment, including both the physical and biotic components.

Economic control: Control of pests such that expenditure on control measures is more than compensated by increased value of yield.

Economic damage: The injury to a crop that justifies the cost of artificial control measures.

Economic entomology: That branch of insect study directed towards preventing man's losses and increasing his gains through manipulation of insect populations.

Economic injury level: The lowest population density of a pest that causes economic damage; the pest density that causes damage equal to the cost of preventing the damage.

Economic pest: A pest causing a crop loss of about 5-10%.

Economic poison: Any substance or mixture of substance intended for preventing, destroying, repelling or mitigating any insects, rodents, nematodes, fungi or weeds or any other form of life declared to be a pest and any substance or mixture of substances intended for use as a plant regulator, defoliant or desiccant.

Economic threshold: 1. A population density concept which allows the determination of the point at which pest numbers are sufficient to cause economic injury unless suppressive action is taken. 2. That point of pest infection where application of a control measure would return more money than the cost of the control procedure. 3. Pest population level at which control measures should be started to prevent the pest population from reaching economic injury level.

Ecosystem: 1. Interacting system of living organisms in an area and their physical environment. 2. The functional ecological system of a given area comprised of the community of plant and animal populations and the non-living environment.

Ecto-type. A small group of plants or animals locally adopted to a particular habitat.

Ectoderm: The embryological layer which gives rise to the nervous system, integument and several other parts of an insect.

Ectognathous (**Ectognathy**): Insects with exerted mouth parts.

Ectohormone: A substance secreted by an animal to the outside causing a specific reaction, such as determination of physiological development, in a receiving individual of the same species.

Ectoparasites: Plants and animals that live and feed on the outside of an animal or plant. Most are annoying, cause injury, or carry disease organisms. Examples lice, fleas, some fungi.

Ectoparasitoid: An insect parasite which develops externally on its arthropod host.

Efferent nerve: A nerve that conducts from a nerve center toward the periphery; the axon of a motor neurone.

Efficacy: Capacity for serving to produce effects - effectiveness.

Effluent: Liquid flowing out. The term is generally used for unwanted liquid wastes from factories, industries and city drains.

Ejaculatory duct: Tube through which the spermatozoa pass from the vasa differentia.

Elateriform larva: A larva with the form of a wireworm; that is long, slender, heavily sclerotized, with short thoracic legs, and with few body hairs.

Electron microscope: A microscope which works on electronic beam penetrating the thin film of the material (object) producing a much magnified image on a fluorescent screen which is recorded by a camera. The resolving power of an electron microscope is much greater than an ordinary microscope.

Elytra (sing:, elytron): The fore wings of beetles. They are hard shell-like structures for protecting the membranous hind - wings; wing cases.

Embolium: A narrow part of the corium, separated by a suture, along the costal margin in front wing of Hemiptera.

Emery's rule: The rule that species of social parasites are very similar to their species and therefore presumably closely related to them phylogenetically.

Emigration: Movement of individuals out of a population.

Empodium: A single pad - like or filiform median structure often present between the claws of insects.

Emulsifiable concentrate: 1. A liquid formulation, usually containing a pesticide, which will form an emulsion when mixed with water or certain other liquids. 2. Produced by dissolving the toxicant and an emulsifying agent in an organic solvent. A solvent substantially insoluble in water is usually selected since water miscible solvents have not in general proved satisfactory. Strength usually stated in pounds of toxicant per gallon of concentrate.

3. Liquid formulation that immediately disperses as fine droplets in water to form an emulsion. 4. A liquid formulation of insecticide which contains an emulsifier so that water may be added to form an emulsion.

Emulsifying agent: 1. Also emulsifier. Substance when added to emulsion of insecticide helps resist tendency of oil droplets and other water unmiscible liquids to coalesce when mixed with water and thus help stabilize emulsion. Used for making stock emulsions. 2. A material which facilitates the suspending of one liquid in another; for example, oil dispersed in water.

Emulsion: 1. A finely divided, fatty or resinous substance held in suspension in an aqueous liquid. 2. A mixture of liquids that do not dissolve in each other. 3. Stable dispersion of oil droplets in aqueous solution.

Encapsulation: 1. A method of formulating pesticides, is which the active ingredient in encased in a material (often poly vinyl) resulting in sustained pesticidal release and decreased hazard. Also a method of disposal of pesticides and pesticide containers by sealing them in sturdy, waterproof, chemical - proof containers which are then sealed in thick plastic, steel, or concrete to resist damage of breakage so the contents cannot get out. 2. Or microencapsulation: the encapsulation of a pesticide in a non-volatile envelope of gelatin, usually of minute size for delayed release.

End chamber: The germarium of a gonadial tube.

Endite: A mesal lobe of any limb segment.

Endocrine: Secreting internally, applied to organs whose function is to secrete into blood or lymph a substance which has an important role in metabolism.

Endocuticle: It is a thicker, softer and flexible innermost layer of the cuticle.

Endoderm: The innermost layer of cells produced in an early embryo that becomes part of the lining of the alimentary canal.

Endoparasite: A parasite living within the body of the host.

Endoparasitoid: An insect parasite which develops within the body of its arthropod host.

Endophallus: The inner chamber of the phallus invaginated at the end of the aedeagus, into which the ejaculatory duct opens; typically an eversible sac or tube, but sometimes a permanently internal phallic structure.

Endopodite: The mesal branch of a biramous appendage; the main shaft of the limb beyond the basipodite.

Endoskeleton: 1. It is the internal framework of rigid processes of the body wall. Such a framework in the head is called tentorium. 2. Chitinous processes extending in ward into the body cavity from the body wall for attachment of muscles.

Enteric caeca, gastric caeca (sing:, caecum): Sac-like diverticula at the anterior end of the mid - gut.

Entognathous (entognathy): Mouthparts hidden in the head.

Entomogenous: Growing in or on an insect, for example certain fungi.

Entomophagous: An animal (or plant) which feeds upon insects.

Entomopox virus: An insect virus which bears great morphological similarity to the poxviruses of vertebrates. They are characterized by spindle shaped to oval virus particles, surrounded by 2-3 limiting membranes and occluded in large oval protein - aceous crystals formed in the cytoplasm of infected cells.

Entoparasite: An internal parasite.

Environment: 1. The total surroundings of an organism. 2. Means water, air, land, all plants, man and other animals living therein, and the interrelationships which exist among them.

Environmental resistance: The sum total of physical and biotic mortality processes in the environment of an organism which prevent it from expressing its absolute reproductive potential.

Enzyme: 1. A natural substance which regulates the rate of a reaction but which itself remains chemically unchanged. 2. An organic catalyst formed by a living cell. 3. A protein that acts as a catalyst in the chemical reactions of living organisms.

EPA: Environmental Protection Agency. Responsible for the protection of the environment in the united state.

Epicranial suture: The dorsal Y-shaped suture of the cranium, including the median coronal suture (cs) of the vertex, and the divergent frontal sutures (fs) of the facial region.

Epicranium: A term variously applied to the entire cranium, to the cranium exclusive of the frons, or preferably to the upper part of the cranium.

Epicuticle: It is an outer film-like water proof layer of the cuticle.

Epidemic: A sudden widespread increase in the incidence of a disease or organism.

Epidermis: The cellular layer of the body wall that secretes a thick cuticle on its outer surface. It is made up of epithelial cells.

Epimeron (pl: epimera): The posterior division of a thoracic pleuron.

Epimorphic development: Immatures possess same number of body segments as adults.

Epiphallus: A sclerite in some Orthoptera in the floor of the genital chamber proximal to the base of the phallus. (Pseudosternite.)

Epipharynx: A mouth - part structure attached to the inner surface of the upper lip. Sometimes it is elongated to form a part of the proboscis.

Epiphysis: A mobile pad or lappet like process on the inner aspect of the front tibia of some Lepidoptera.

Epipleuron (pl: epipleura): The diflexed or inflexed portion of the elytron, immediately beneath the edge.

Epiproct: The dorsal part of the eleventh segment of the abdomen

Episternum: The anterior sclerite of the pleuron of the thorax.

Epistomal suture: The frontoclypeal suture; a groove uniting the anterior ends of the subgenal sutures across the face, forming internally a strong epistomal ridge (ER), typically straight, but often arched upward, sometimes absent.

Epistome: The part of the face just above the mouth; the oral margin (Diptera).

Epithelial: Pertaining to or of epithelium.

Epithelium: The layer of cells that covers a surface or lines a cavity.

Eradication: The complete elimination of either weed, insects, disease organisms or other pests from an area.

Ergatogyne: Any form morphologically intermediate between the worker and the queen.

Eri silkworm: *Attacus ricinis.* Domesticate silk worm feeding on castor leaves. Reared in Assam and northern West Bengal. Produces eri silk.

Eruciform: Resembling a caterpillar, with cylindrical body and with both thoracic legs and prolegs.

Eruciform larva: A caterpillar, a larva with cylindrical body, well developed head, thoracic legs, and abdominal prolegs.

Esophagus: The narrow part of the alimentary canal immediately posterior to the pharynx and mouth

Estivation: Reduction of biological activity by an organism in response to periods of hot or dry weather.

Eucone eye: Eyes in which the cone cells are almost completely transformed into a crystalline cone.

Euphagous: Feeding on specific host plant.

Eusternum: The intrasegmental ventral plate of a thoracic segment, exclusive of the spinasternum, but usually including the sternopleurites.

Evagination: An out-pocketing or saclike structure on the outside.

Exarate pupa: A pupa which has its appendages free and not glued to the body.

Excretion: The elimination of waste products of metabolism.

Exocuticle: The hard and usually darkened layer of the cuticle lying between endocuticle and epicuticle.

Exopodite: The layer of the branch of a biramous appendage.

Exoskeleton: The segmented, external skeleton of an insect; the insect's "skin".

Exotic: Not native to a place or country but introduced from foreign land.

Exotoxin: A soluble toxic substance produced by certain bacteria and found in their surrounding growth medium; Also called endo-toxin.

Exposure: When contact occurs with a pesticide through skin (dermal), mouth (oral), lungs (inhalation/respiratory), or eyes.

External genitalia: The reproductive structures concerned with mating and deposition of eggs.

Exuvia: The cast skin of a larva or nymph. This term does not exist in the singular.

Exuvial glands: Glands of the epidermis supposed to secrete the exuvial or molting liquid.

Exuvium (pl: exuviae): Cast off skin after molting

Facet: The lens which forms the base of the visual element, the ommatidium, visible on the surface of the compound eye.

Factitious host: An unnatural, but acceptable host used in laboratory propagation of beneficial organisms.

Facultative: Incidental, not necessarily compelled to live under one type of environment.

Facultative parasitism: Here in reference to nematodes which may either parasitize healthy insects, or develop in some other way in the environment (e.g. mycetophagy) if no insect host is encountered.

Fat body: An organ in the insect body with multiple functions in metabolism, food storage, and excretion. "Fat body" is a misnomer, for protein and glycogen are stored as well as fat.

Fauna: Assemblage of animal life in certain environment or place.

Feces: Indigestible wastes discharged from the digestive tract.

Fecundity: 1. Capacity to produce off springs. 2. In insect ecology, the number of eggs per female that hatch or become larvae. 3. The word is used sometimes for expressing fertility in agronomy i.e. the number of plants or grain yield per square unit area of soil.

Femur (pl:, femora): The 3rd stout part of the leg which is articulated to the body through trochanter and coxa.

Fiducial limits: The lower and the upper values for the estimate of LD_{50} or LD_{90} computed from the probit regression equation at a given reliability coefficient, say 95% or 99%.

Filiform: Thread-like or hair-like; that type of- antenna with the segments quite uniform in size.

Filter chamber: A part of the alimentary canal in Homoptera in which the two ends of the ventriculus and the beginning of the intestine are bound together in a membranous and muscular sheath.

Flagellum: The part of the antenna distal to the pedicel, typically filamentous, but of various forms, usually subsegmented or multi articulate.

Flame thrower: Compressed air type of sprayer filled with kerosene oil for producing flame. Lance modified to carry burner which is heated before allowing oil to flow through. Used for burning settled swarms of locusts and scrub weeds.

Fog or smoke generator: Machine which disperses spray material in extremely fine droplets (1 to 50 microns in diameter) which remain air-borne for long time. Successfully used in dense foliage or closed space; covers object uniformly with chemical.

Follicle cells: The inner epithelial cells of an ovarian egg tube.

Food chain: 1. The direction of energy flow between species that feed upon one another. 2. All assemblage of organisms, including producers, consumers, and decomposers, through which energy and materials may move in a community.

Fore wing: The anterior pair of wings in insects.

Foregut: The anterior portion of the alimentary tract, from the mouth to the midgut.

Formulation: 1. Statement of nature and amount of all constituents of a pesticide concentrate. 2. Method of preparation of a pesticide concentrate. 3. The pesticide product containing the active ingredient, the carrier and other additives required to make it ready for sale. 4. The type of preparation in which a pesticide is sold for use. 5. The preparation of the active ingredient in a form suitable for use is referred to as formulation.

Fossil: A relic or trace of a former living thing preserved in the rocks.

Fossorilal legs: Legs adapted for digging or burrowing.

Frenulum: The spine or spines which arise from the base of the hind wing in many Lepidoptera and project beneath the fore wing.

Frons or front: That region of the head lying between the eyes and the clypeus.

Frontal ganglion: The median precerebral ganglion of the stomodaeal system.

Frontal suture: The crescentic suture cutting across the frons above the base of the antennae (in Diptera) and usually extending downward on each side to near the mouth.

Fuller's earth: A silicate mineral used as an adsorbent base or dust diluent. It is flowable and possesses excellent grind ability acting as a grinding aid.

Fumigation: The use of poisonous gases for the destruction of pests; mainly used against insects and rodent.

Furuca: The forked endosternal process of higher insects, formed of the sternal apophyses supported on a median inflection of the sternum.

Furcasternum: A distinct part of the sternum in some insects bearing the furca. (The term generally applied to the sternelum.)

Furcula: A springing organ on the ventral surface of the abdomen of Collembola.

Galea: The outer endite lobe of a maxilla, provided with a muscle arising in the stipes.

Gall: 1. Bile, secretion of liver. 2. A plant tumour found practically on any part of the plant; caused by irritation from fungus, bacterium or insect.

Gamete: A sexual reproductive cell that must usually fuse with another before development occurs; an egg or sperm.

Gametocyst: A sex cell stage of plasmodium, the malarial parasite.

Gamma irradiation: The use of rays from a radioactive source.

Ganglion (pl: ganglia): Nerve center containing a cell mass and fibers; white disc-like bodies connected by a double cord.

Gastric caeca: The sac-like diverticula at the anterior end of the midgut.

Gastrulation: The formation of the endoderm, either by invagination of one wall of the blastula or by internal proliferation of cells from the blastoderm.

Genae: The lateral parts of the parietals, generally the areas behind and beneath the eyes.

Gene: A unit of inheritance.

Gene pool: The sum total of all the genes of all the individuals in a population.

Geniculate antennae: Elbowed type of antennae.

Genital chamber: In the female a copulatory invagination cavity behind or above the eighth abdominal sternum containing the gonopore and the orifice of the spermatheca, often narrowed to form a pouch like or tubular vagina. (Bursa copulatrix.)

Genital claspers: Organs of the male genitalia which serve to hold the female during copulation.

Genital ridge: One of the embryonic gonadial rudiments, a ridge like swelling of the splanchnopleure wall of the mesoderm containing the germ cells.

Genital segment: Specifically the ninth segment of the abdomen in the male, though other segments are frequently associated with the ninth in the genital complex. (Gonosomite.)

Genitalia: 1. The organs of reproduction collectively, but usually applied only to the external genitalia. 2. A term applied collectively to all the genital structures. 3. The reproductive organs; the external which enable the sexes to copulate and the females to deposit eggs, strictly these are the external genitalia.

Genotype: The particular combination of genes present in the cells of an individual.

Germ band: The area of thickened cells on the ventral side of the blastoderm that becomes the embryo. (Embryonic rudiment, germ disc, primitive streak, Keimstreif, Keimscheibe, bandclctte, plaque ventrale, piastra germinativa.)

Germ cells: Reproductive cells, or more specifically the early undifferentiated reproductive cells designed to become ova or spermatozoa.

Germarium: The end chamber of an ovarial or testicular tube, containing the primary oogonia or spermatogonia. (Endkammer.)

Germicide: A chemical or agent that kills microorganisms such as bacteria or prevent them from causing disease.

Germination: Process of germinating or beginning of vegetative growth. Often refer to the beginning of growth from a seed.

Gizzard, proventriculus: A part of the fore-gut immediately behind the crop. It is poorly developed in sucking insects. In chewing-insects, it is equipped with, powerful muscles and teeth with which the harder particles of food can be crushed.

Glossa: The inner part of the extremity of the labium of insect mouth parts.

Glutathione conjugation: This conjugation of herbicides and insecticides is a common reaction resulting in attack of alkyl or aryl halide at the sulphur atom of the glutathione to give S-alkyl or S-aryl glutathione derivatives. Stable complexes with glutathione and triazine herbicides have been isolated from plants.

Glycogen: A carbohydrate synthesized by animals, also called "animal starch."

Gnathal: Pertaining to jaws or man dibles.

Gnathal segments: The segments of the insect embryo the appendages of which become the mandibles and first and second maxillae.

Gonad: The ovary or testis or their embryonic rudiments, formed by splanchnic mesodermal cells covering the germ cells.

Gonoduct: Duct of gonad to the genital opening in Female, the oviduct and in male, the vasa defrentia.

Gonopore: In the male the external opening of the median ejaculatory duct, usually concealed in the endophallus, or one of the apertures of paired exit ducts.

Granular formulation: 1. A product in the granule form, usually available in various mesh sizes e.g. 15/30, 20/35, 20/40, 24/40, 30/60 etc., required size being chosen on the basis of insect population, type of crop and condition of cropped area. These are further classified as extruded granules (which readily disintegrate in water) and non-extruded granules (which will not disintegrate in water before a certain lapse of time). Such formulations are specified with a suffix G after the concentration of active ingredient present in them e.g. Phorate-10 G means a granular formulation of phorate having an a.i. 10% w/w. 2. A formulation of dry ready-to-use low-concentrate pesticide plus an inert carrier.

Granulosis: An insect viral disease characterized by the presence of minute granular inclusions (capsules) in the infected cells.

Gravid: An insect containing fertilized eggs.

Grinding: Cutting with molars or dentes molars.

Grooming: The licking of the body surfaces of nestmates. Self-grooming also occurs, in which individuals clean their own bodies both by licking and stroking with the legs.

Growth regulator: An organic sub-stance effective in minute amounts for controlling or modifying plant processes; organic compounds, other than nutrients, which in small amounts promote, inhibit, or otherwise modify any physiological process in plants.

Grub: A term applied in the general sense to insect larvae such as caterpillars and legless maggots, but sometimes used more specifically for the legged larvae of beetles.

Gula: A median ventral plate of the head of some insects, developed as a sclerotization of the gular region of the neck proximal to the posterior tentorial pits, continuous with the basal plate of the labium.

Gular sutures: The ventral ends of the postoccipital suture extended forward on the underside of the head in some prognathous insects.

Gynandromorph: An individual containing patches of both male and female tissue.

Habitat: Physical place where an organism lives.

Haemocoele: The blood cavity or cavities of the embryo between the mesoderm and the other germ layers, probably a remnant of the blastocoele.

Haemocytes: The free floating cells of the blood or circulatory fluid.

Haemoglobin: The oxygen carrying red pigment of the blood. It is absent in most of the insects.

Haemolymph: The unpigmented blood of insect; the fluid in the body of some coelomates.

Half life: Time taken for half the deposit of a pesticide chemical to disappear or degrade.

Haltere or halter (pl: halteres): Small, knobbed organ, one on each side of the metathorax, representing the hind wing of the Oiptera.

Hamuli (sing: hamulus): Minute hooks; a series of minute hooks on the anterior margin of the hind wing, with which the front and hind wings are attached together in Hymenoptera.

Haploid: Possessing a single set of unpaired chromosomes (n).

Haustellate: formed for sucking the mandibles wanting or not for chewing.

Haustellum: The distal portion of the proboscis of some dipterous insects, adapted for sucking exposed liquids.

Head capsule: The welded together sclerites of the head forming a hard compact mass.

Hemimetabola: Insects with simple metamorphosis, with immature stages aquatic and adults terrestrial; insects of the orders Odonata, Ephemeroptera. andPlecoptera; young are called naiads.

Hemocoel: The body cavity which is normally filled with blood orhemolymph.

Herbicide: A pesticide used for killing or inhibiting plant growth. A weed or grass killer.

Herbivorous: Feeding or living on grass or herbage.

Hermaphrodite: An individual, such as cottony cushion scale, *Icerya purchasi,* which contains both the male and the female reproductive organs. The individual is self fertilizing.

Heterometabola: Insects with simple metamorphosis including the paurometabola and hemimetabola.

Hexapoda: 1. The true insects. 2. Insects are also called hexapoda owing to the presence of six legs.

Hibernate: To enter a state of dormancy over the winter period.

Hibernation: A period of suspended animation in animals during seasonal low temperatures.

High volume sprays: Spray applications of more than 50 gallons per acre.

Highly toxic: 1. substances are considered highly toxic by law if the LD_{50} of a single oral dose is 50 milligrams or less per kilogram of body weight (LD_{50}); 2. If LC_{50} o of toxicity by inhalation is 2,000 mcg of a gas or vapour when administered by continuous inhalation for one hour to both male and female rats or to other rodent or non - rodent species if it is reasonably forseeable that such concentrations will be encountered by man or: 3. If LD_{50} of toxicity by skin absorption is 200 milligrams or less per kilogram of body weight when administered by continuous contact for 24 hours with the bare skin of rabbits or other rodent or non - rodent species as specified.

Hind-wings: Wings developed on the third thoracic segment.

Hind-gut: The posterior portion of the alimentary tract, between the mid gut and anus.

Histopathology: The study of abnormal microscopic changes in the tissue structure of diseased organism

Holometabola: Insects having complex metamorphosis. The pupal stage is present.

Holometabolous: Undergoing a complete metamorphosis during development, with distinct larval pupal, and adult stages, the Hymenoptera, for example are holometabolous (opposed to hemimetabolous).

Holopneustic: With all the spiracles open and functional.

Homeostasis: The maintenance of a steady state, especially a physiological or social steady state, by means of self regulation through internal feedback responses.

Homonomous: Similar in form, function or development.

Homonym: One or the same name for two or more different things.

Honeycomb: The waxy structure in which honey is stored by bee.

Honeydew: 1. Certain insects, such as aphids and coccids, owing to almost continuous feeding on the juice of plants, take in more sugar and water than they need. The excess excreted as sweetish excretion is called honey dew; sweet thick fluid. 2. Sweet and sticky fluid discharged from the anus (not cornicles) by some Homoptera.

Hormone: 1. A naturally occurring substance in plants or animals that controls growth or other physiological processes. it is used with reference to certain man made or synthetic chemicals that regulate or affect growth activity. 2. An internally secreted chemical messenger substance, produced by endocrine tissues (glands), which influences or activates other organs or physiological processes within the same organism: see Pheromone. 3. A chemical secreted in one part of the body that regulates the function of some other part of the body.

Hornet: A large wasp of the family Vespinae, particularly a member of the genus vespa or (in the United States) the bald-faced hornet Vespula (Dolichovespula) maculata.

Hoses: These are tubes that are used in the sprayer.

Host: 1. The term "host" means any plant or animal on or in which another lives for nourishment, or protection. 2. Organism in on which an insect feeds; the organism may be plant (pest) or animal (parasite).

Host resistance: Here in the sence of Painter (1951) meaning the relative genetic ability of a certain variety of plant (or animal) to produce a larger crop of good quality than do ordinary varieties at the same level of pest infestation: practically speaking, based on three elements: non preference, antibiosis, and/or tolerance.

Humeral: Pertaining to the shoulder; located in the anterior basal portion of the wing.

Humeral angle: The basal anterior angle or portion of the wing.

Humeral bristles: Bristles on the humeral calius (Diptera).

Humeral cross vein: A cross vein in the humeral portion of the wing between the costa and the subcosta.

Humeral plate: The anterior preaxillary sclerite of the base supporting the costal vein; very large in Odonata.

Humeral vein: A branch of the sub costa which serves to strengthen the humeral angle of the wing (Neuroptera, Lepidopetra).

Humerus (pl:, humeri): The shoulder; the lateral angle of the pronotum (Hemiptera).

Hyper parasite: A parasite which attacks another insect which is itself a parasite.

Hyperestrogenic syndrome: A condition in which the esterogenic hormone is present in undue high proportion.

Hypermetamorphosis: 1. A type of complete metamorphosis in which the different larval instars represent two or more different types of larvae. 2. A type of parasitic insect life cycle involving transformation through at least two distinctly different larval types, especially an active host-seeking first instar, and several more passive parasitic instars.

Hyperparasite: A parasite whose host is also a parasite.

Hypodermis: The cellular layer of the body wall which secretes the cuticula.

Hypognathus: With the head vertical and the mouth parts located ventrally.

Hypomere: A ventral process of the phallobase.

Hypopharynx: 1. Sensitive and sensory structure on the upper surface of labium serving as organ of taste or true tongue. 2. A tongue-like lobe in the mouth. It is elongated and forms a part of proboscis in blood sucking insects, such as mosquitoes.

Hypopleural bristles: A more or less vertical row of bristles on the hypopleuron, usually directly above the hind coxa (Diptera).

Hypostomal suture: The part of the subgenal suture posterior to the mandible, often obsolete or suppressed.

Hypostome: In ticks, the median ventral dart-like mouthpart that is immovably attached to basal part of the capitulum.

Ileum: The anterior part of the anterior intestine, between the ventriculus or pylorus and the colon. (Small intestine, Dunndarm).

Imaginal instar: Pertaining to adult stage or imago.

Imago: 1. The last or adult stage in insect metamorphosis; the perfect insect. 2. The adult insect. In termites, the term is usually applied only to adult primary reproductives.

Immature stage: Any stage of an insect prior to the imago or adult form.

Immigration: Movement of individuals into a population. Immune. 1. Exempt from infection. 2. A state of not being affected by disease or poison; exempt from or protected against. Incidence (of disease, fungal or insect attack). The number of new cases of a particular malady within a given period of time (season, year) in a population under study.

Inclusion body: The proteinaceous or crystal-like structure produced in insect cells infected with certain viral pathogens; it occurs in various shapes and size, and usually encloses a number of replicated virions.

Incompatibility: The inability of a pesticide to mix with other pesticides without producing undesirable effects.

Incompatible: Not capable of being mixed or used together; in the case of pesticides, the effectiveness of one or more is reduced; or they cause injury to plants or animals.

Incomplete metamorphosis: 1. Insect metamorphosis in which the young are hatched in general adult form and develop without the quiescent stage. 2. Hemimetabolic development; the immature (nymphs) do not resemble the adults and a pupal stage is lacking.

Indigenous: Native to a particular region or country.

Indirect pest: An organism which causes delayed or insidious degradation of marketable produce through generalized attacks on the host which may reduce growth, vigour or yield.

Induced resistance: Resistance of a plant to attack of pest or disease brought about by conditions under which it is growing, and varying in degree according to these conditions; generally brought about by farm practices and nutritional means.

Inert: A material having no biological activity, such as diatomaceous earth, silica aerogels, magnesium oxide, aluminum oxide, activated charcoal, cow-dung ash and soap stone.

Infect: To enter and establish a pathogenic relationship with a plant (host); to enter and persist in a carrier.

Infection: The development and establishment of a pathogen (e.g., a bacterium) in its host which will produce a disease.

Inferior appendage: The lower one (Anisoptera) or two (Zygoptera) of the terminal abdominal appendages used in grasping the female at the time of copulation (male Odonata).

Infest: 1. To occupy and cause injury to either a plant or to soil, or stored products. 2. To be present in number (e.g., insects, mites, nematodes, bacteria, fungi) on the surface matter or in the soil. Do not confuse with infect (infection) that applies only to living, diseased plants or animals. 3. To beset and cause injury to plant, soil or stored products.

Infestation: Pests that are found in an area or location where they are not wanted.

Ingested: Taken into the digestive system.

Ingluvial: Pertaining to the crop, the distensible middle portion of the foregut in which (in many species) liquid food is stored.

Ingluvial ganglion: A paired ganglion of the stomodaeal nervous system in some insects, situated on the side of the crop. (Gastric ganglion.)

Inhalation: To make air into the lungs; to breathe in.

Inhalation toxicity: Poisoning through the respiratory system.

Inhibitor: A chemical, usually of the regulator type, that prevents or suppresses growth or other physiological processes in plants 2-amino-1, 2, 4-triazole (ATA) is an example of a growth inhibitor.

Inoculation: The introduction of an infective agent (the inoculum) into or on living tissues producing a specific disease.

Inquilinism: The relation in which a socially parasitic species spends the entire life cycle in the nests of its host species. Workers are either lacking or, if present, are usually scarce and degenerate in behavior. This condition is sometimes referred to loosely as "Permanent Parasitism".

Insect growth regulator: A general class of natural and synthetic chemical compounds associated with the control of growth and metamorphosis in insects and including the juvenoids plus several other chemicals that inhibit either JH or other physiological activities.

Insect parasites: Insects which lay eggs on the body of other insects, when the emerging larvae feed internally on them and kill them.

Insect pathology: The study of insect diseases, their causes and treatments.

Insect pest management: An ecologically based strategy of maintaining insect pest population below the economic injury level by the use of any or all control techniques that are economically, ecologically and socially acceptable.

Insect predators: Insects which catch other insects and feed on them.

Insect society: In the strict sense, a colony of eusocial insects (ants, termites, eusocial wasps, or bees).

Insect vector: An insect which carries virus, bacterium or spores or mycelium of a pathogenic fungus and inoculates susceptible plants or animals.

Insect (s): 1. A class of arthropod with bodies divided into a head, thorax and abdomen, the head bearing a pair of feelers or antennae, the thorax three pairs of legs and wings. 2. An air-breathing animal of phylum Arthropoda, with a distinct head, thorax and abdomen; has one pair of antennae, three pairs of legs, usually two pairs of wings. Some infest plants and animals; some are insectivorous, some pollinate plants; some produce edible or otherwise useful products. 3. Any of the numerous small invertebrate animals generally having segmented bodies and for the most part belonging to the class insects, comprising six-legged, usually winged forms.

Insectary: Also called insectarium. A laboratory in which insects are bred in large numbers.

Insecticide: 1. A pesticide which kills insects. Most of those in farm use today are synthetic organic compounds and include chlorinated hydrocarbons, organo-phosphorous compound and carbamates. They are available in granular, liquid and powder forms and may be applied in various ways. 2. Chemical prepared with the express purpose of controlling.

Insect pests: active ingredients can be gas, liquid or solid, formulated into liquefied gases, smokes, dusts, water-dispersible powders, emulsions and solution. 1. A toxic chemical substance employed to kill and control insects. 2. A substance or mixture of substances intended to prevent or destroy any insects which may be present in any environment.

Insecticide resistance: The degree to which a species of insect tolerates a toxic substance. The degree of resistance is based on LD_{50} values of a population that has been previously exposed to insecticide, compared to those of an unexposed population.

Insecticides Act 1968: To minimize pesticide hazard and regulate the use of pesticides in public health and agriculture, Government of India, in 1968, passed a comprehensive legislation called insecticides Act, 1968. However, the Act, along with the rules framed with the Act started functioning (implemented) from 1st August, 1971.Various regulatory provisions made in the act include compulsory registration of pesticides at the Central level; issue of licence for manufacture/formulation and sale at the State level; inter departmental/ministerial/organizational coordination in the various fields of matters arising out of the implementation of this Act; establishment of enforcement machinery like Insecticide Analysts, Insecticide Inspectors etc. under various sections of the Act two statutory bodies namely, the Central Insecticide Board and the Registration committee for pesticides have been constituted.

Insectivorous: Feeding and living on insects.

Instar: 1. A stage of development of an insect between two moults of the hard exoskeleton, but excluding the egg. An adult butterfly is the sixth instar. The first four are caterpillars, of increasing size, followed by a pupa. 2. Any period between molts during the course of development. 3. A period or stage between molts in the larva, first instar is the stage between egg and the first molt.

Integrated control: 1. Control of pests which combines and integrates chemical methods with natural and biological control. Chemical control is applied as necessary and in a manner which is least disruptive to natural and biological control. Also a pest management system that in the context of the associated environment and the population dynamics of the pest species, utilizes all suitable techniques and methods in as compatible a manner as possible and maintains the pest populations at levels below those causing economic injury. 2. Utilizing multiple approaches to pest control, when pesticides are required they should exert the least adverse effects on the pests natural enemies. 3. The control of pest (s) by selecting and utilizing more than one control techniques so that the economic damage is avoided and adverse side-effects on the agro-ecosystem are minimized.

Integrated pest management: A pest management system that, in the context of the associated environment and the population dynamics of the pest species, utilizing all suitable techniques and methods in as compatible a manner as possible to maintain the pest population at levels below those causing economically unacceptable damage to a crop. It is a combination of chemical and biological control (using predators), and utilization of advanced and just cultural practices (including disease or pest resistant varieties).

Integrated pest suppression: An approach to compatible utilization of all available forms of pest suppression, including mechanical, biological, chemical, and natural control, in a systematic fashion, with the primary goal of safe, effective. and economical pest population reduction. It may be directed at a single important pest species by combining a variety of measures against the species or at a complex of pests, integrating the individual protective measures applied against each, so as not to interfere one with the other.

Integument: The outer covering of the body of an insect or other animals.

Interantennal suture: A suture extending between the bases of the two antennae (Siphonaptera).

Intercalary vein: A supplementary longitudinal wing vein lying between two preexisting veins.

Intercostal vein: The subcosta (Hymenoptera).

Intermediate host: The host which harbors the immature stages or the asexual stages of a parasite.

Internal respiration: The process of oxidation accompanying metabolism in the cells of the body tissues.

International unit: An arbitrarily set basis for comparing the efficacy of insect-pathogenic *Bacillus thuringiensis* preparations. An international unit is one one-thousandth of the amount of insecticidal activity contained in one milligram of a preparation of the primary standard E-16 strain of B. thuringiensis, is measured by bioassay against certain lepidopterous larvae (i.e. comparative LD_{50}). A B. *thuringiensis* formulation with a potency of 1000 IU/mg is therefore, equal to the international standard.

A secondary reference standard B. *thuringiensis* strain (HD-1-S-1971) is in use in the USA, and has been assigned a potency of 18000 IU/mg against the cabbage looper. Trichophusia ni (Hubner).

Intersegmental membrane: The flexible conjunctiva between two secondary segments; usually the nonsclerotized posterior part of a primary segment, but variable.

Intersternites (1st): Primary intersegmental sclerites of the venter, becoming the spinasterna of the thorax.

Interstitial: Situated between two segment; (interstitial trochanter of Coeloptera).

Intervalvulae: Sternal sclerites in the venter of the ninth abdominal segment between the second valvifers.

Intestine: The proctodaeum, or the part of the proctodaeum beyond the pylorus (The term applied also to the entire alimentary canal).

Intima: The extension of the exoskeleton into the internal cavities of the body for lining the foregut, hindgut, and tracheae.

Intra-alar bristles: A row of two or three bristles situated on the mesonotum above the wing base, between the dorsocentral and the supra-alar bristles (Diptera).

Intrinsic articulation: Articulation in which the points making contact are sclerotic prolongation in the articular membrane.

Inundative release: A method of periodic introduction of biotic agents which is analogous to insecticide treatment in that a greater amount of the liberated material is used than is actually effective, repetition may be necessary, and the effect is more or less immediate.

Invagination: Pouch or sac occurred due to the infolding or indrawing of the outer surface.

Inversion: A form of mutation in which part of a chromosome becomes ditached from the rest and then joins up again in the oppositic way round.

Inviscate: 1. To trap insects in a sticky substance or on a sticky surface. 2. To make a substance more sticky by mixing it with viscous matter.

IOBC: International Organization for Biological Control of Noxious Animals and Plants (an affiliate of the International Union of Biological Sciences). A global organization of governmental units and individuals interested in biological pest suppression, headquartered in Zurich. Switzerland. Major objectives include dissemination of information coordination, and promotion of research and application of biological pest suppression. Publisher of the journal Entomophaga.

Iridescent virus: A non-occuded insect virus characterized particularly by its fascinating optical properties; thes stem form Bragg reflection, and give purified pellets of virus particles an opalescent appearance.

Iris: Dark pigment surrounding the dioptric apparatus of an eye.

Isomer: 1. Any of the two or more chemical compounds having like constituent atoms, but differing in physical or chemical properties owing to differences in the arrangement of the atoms. 2. Two or more chemical compounds having the same structure but different properties.

Johnston's organ: The sense organ similar to a chordotonal organ, located in the second antennal segment of most insects.

Joint: An articulation of two successive segments or parts.

Jugal region: A posterior basal lobe or area of the wing set off from the vannal region by the plica jugalis, containing the vena arcuata and vena cardinalis when these veins are present.

Jugum: A finger like process extending from the base of the front wing underneath the hind wing (Lepidoptera).

Juvabione ("paper factor"): A JH- active chemical found in paper products manufactured from American balsam fir trees: active against the European red bug, *Pyrrhocoris apterus* L.

Juvenile hormone (JH): 1. A chemical produced by the corpora allata, one of the three major insect developmental hormones, and responsible for determining the type of molt which will occur when one is directed by ecdysone. At high JH blood titer, additional larval or nymphal molts occur; low titer or complete absence of JH causes the pupal or imaginal transformation to occur. 2. The hormone, secreted by the corpora allata, that maintains the immature form of an insect during early molts.

Juvenile stage: Any or every post embryonic stage in the development of an organism that precedes the sexually mature adult stage e.g. egg, larval and pupal stages of insects.

Juvenoid (juvenile hormone analog: Any synthetic or natural chemical of plant or animal origin exhibiting biological activity similar to the true juvenile hormone.

Kairomone: A trans-specific chemical messenger, the adaptive benefit of which falls on the recipient rather than the emitter. see Allomone.

KD_{50}: Fifty per cent knockdown effect on insects.

Keratin: A progein forming the principal matter of hair, horns, and nails.

Key factor: The hypothetical causative agent most responsible for population density change. In practice, a variable factor in the environment recognized as most closely linked to changes in population density, and measurable by continuous monitoring for use in predicting future population trends (Morris. 1959, 1963).

Key pest: An insect pest or disease normally present at some time during the growing season that causes economic damage to a crop.

Knock down effect: The rapid falling down of the insects treated with an insecticide.

Labellum (pl: labella): The modified tip of the labium of a number of Diptera.

Labial glands: The usual "salivary glands" of insects, opening by median duct between the base of the hypopharynx and the labium, or on the hypopharynx.

Labial palpus (pl:labial palpi): One of the pair of sensory appendages (feeler like and two to five segmented), tile insect labium.

Labial suture: The suture of the labium between the prementum and the postmentum, always distal to the mentum when the latter is present.

Labiostipites (Lst): The prementum, or that part of the labium formed by the stipites of the component labial (second maxillary) appendages.

Labium: The lower "lip," or lowermost mouth part-bearing segment of insects, located just below the mandibles and the maxillae.

Labrum: Sclerite beneath the clypeus; serves as upper lip in insects that have chewing mouth parts.

Labrum epipharynx: A mouth part composed of the labrum and epipharynx and usually elongate.

Lacinia: The inner endite lobe of a maxilla, provided with a muscle arising in the stipes, and often with a second muscle arising on the cranial wall.

Lamellate: Plate-like or sheet-like; composed of or covered with thin sheets.

Lamellate antenna: Antenna having plate-like structures at its distal end.

Lance: An accessory in sprayers; a metal tube of uniform cross section on one end of which the spray nozzle is mounted while the other end is connected to the liquid container with a hose pipe or pressure tubing.

Lanceolate: Lance-shaped; widening above the base and gradually tapering upwards to a point; much longer than broad.

Larva. 1. An immature stage which is radically different in form from the adult; characteristic of the holometabolous insects, including the hymenoptera. In the termites, the term is used in a special sense to designate an immature individual without any external trace of wing buds or soldier characteristics. 2. Young insect that comes out of an egg in early stage of morphological development and differs fundamentally in form from the adult. **Protopod:** Primary larva, little more than prematurly hatchembryo, devoid of abdominal segmentation and having rudimentary cephalic and thoracic appendages (parasitic hymenoptera). **Polypod** (Eruciform): Embryo fully developed, complete segmentation of abdomen and full complement of appendages (Lepidoptera, sawflies).**Oligopod (Campodiforrn):** Embryo hatched in advance stage of development; bear a general resemblance to adult (Thysanoptera). **Apodous:** Derived from oligopod, its trunk appendages are completely suppressed.

Larvicide: An agent which destroys insect larvae.

Lateral movement: Chemical movement in a plant or in the soil to the side or horizontal movement in the roots or soil layers.

Lateral nerve cords: The lateral strands of nerve tissue produced from the ventral neuroblasts.

Lateral ocellus: The simple eye in holometabolus larvae. Also called stemma (pl. stemmata).

Lateral oviduct: In insect, one of the paired lateral ducts of the female genital system connected with the ovary.

Lateral plates: The lateral areas of the germ band after differentiation of the middle plate.

Lateral tracheal trunk: The usual longitudinal tracheal trunk on each side of the body closely connected with the lateral spiracles.

Laterosternite: The lateral part of a definitive thoracic sternum apparently derived from ventral arc (sternopleurite) of the subcoxa.

Laterotergite: A lateral sclerotization of the dorsum distinct from a principal median tergite (Paratergite).

LC_{50}: A means of expressing the toxicity of a compound present in air as a dust, mist, gas or vapour. It is generally expressed as micrograms per liter as a dust or mist but in the case of a gas or vapour as ppm. The LC_{50} is the statistical estimate of the dosage necessary to kill 50 percent of a very large population of the test species, through toxicity on inhalation under stated condition or by law, the concentration which is expected to cause in 50 percent of the test animals so treated.

LD_{50}: A lethal dose for 50 percent of the test organisms. The dose of a toxicant producing 50 percent mortality in a population. A value used in presenting mammalian toxicity, usually oral toxicity, expressed as mg of toxicant per kg of bodyweight.

Leg: Organ of locomotion used in walking.

Legal control: Control of pests through legislation that enforces control measures or imposes regulations, such as quarantines, to prevent the introduction or spread of pests.

Legal residue: Residue that is within safe levels according to the regulations.

Lethal: Fatal or deadly.

Leucocyte: A white blood corpuscle containing no respiratory pigment. They are classified as granular and non granular and as basophil or audophill.

Life cycle: Definite pattern of birth, growth, reproduction, old age and death.

Life history: Habits and changes undergone by an organism from the egg stage to its death as an adult.

Life system: A population and the environment which influences it.

Life table: A device for expressing in an orderly fashion, observations on the changing density of an insect population in time and space, and the processes which direct those changes, especially in relation to the age specific distribution of mortality and its causes.

Ligula: The terminal lobes of the labium collectively, or a terminal part of the labium formed by the union of the lobes.

Lipophilic: Fat loving, used to describe a molecule or group of atoms tending to promote oil-rather than water-solubility. Contrast hydrophilic.

Lipophobic: Used as opposite of lipophilic and therefore also of hydrophobic.

Longevity: Life span of an individual; long life.

LT_{50}: The time required for a toxic substance to kill 50 percent of a test population.

Maggot: 1. The larva of a fly. It is without legs and without well developed head capsule. The head end is pointed and the posterior end appears as if cut off at the tip. 2. Vermiform larva, devoid of true- legs and without a distinct head capsule (Diptera). There are five distinct forms of parasitic diptea. **Caudate:** having paired caudal appendages. **Microtype:** small sized without extensive cuticular armature and mouth parts greatly reduced. **Muscoidiform:** plain maggot without adoptive characters. **Planidium:** more or less ellongated form. often highly sclerotized and bearing heavy spines. **Vesiculate:** the vesicle is in form of a large plate occupying main portion of ventral face of last segment.

Man-made pest: A species which is a pest only because of human interference with natural control processes that normally regulate its density to non-pest levels. Most commonly generated through pest upsets, i.e. the unintended destruction of the natural enemies of a non-pest species with a chemicalinsecticide, or through monocultural practices.

Mandibles: Appendages of the fourth body segment that become the first pair of mouthparts in mandibulate species.

Mandibular glands: A pair of glands often present in insects opening mesally at the bases of the mandibles.

Mandibulate: With jaws suited for chewing mouth parts derived from mandibles.

Mandibulate soldier: A soldier which has mandibles used in colony defense.

Marginal cell: A cell in the distal part of the wing bordering the costal margin (Diptera, Hymenoptera).

Marginal vein: A vein on or just posterior to the wing margin; the vein forming the posterior side of the marginal cell (Hymenoptera).

Maxilla (pl: maxillae): 1. One of the paired mouth part structures immediately posterior to tile mandibles. 2. Upper jaw; part of upper jaw behind premaxilla. In insects and arthropods, the two acccessory jaws or appendages immediately behind the mandibles.

Maxillary glands: Glands present in some insects opening mesally at the bases of the maxillae.

Maxillary palpus (pl: maxillary palpi): A small feeler like structure arising from the maxilla.

Maximum dosage: The largest amount of a pesticide that is safe to use without resulting in excess residues or damage to whatever is being protected.

Maximum residue limit: Amount of pesticide residue permitted to remain in a food. Expressed as parts per million (ppm).

Mechanical control: Control of pests by mechanical means such as window screens, earth barriers, and so on.

Median cell: A cell between the subcostal and median veins at the base of the wing (Hymenoptera).

Median effective dose (ED_{50}): The concentration of a chemical (dose) that will produce a response in half the test subjects. It is an indirect measure of the mean tolerance of a batch of the test subjects. The median lethal dose (LD_{50}) is a special case in which death is the response.

Median effective time (ET_{50}): The time at which a response occurs in half the text subjects after exposure to a pathogenic (including toxicological) stimulus.

Median lethal dose of radiation, (M) LD_{50}: The dose of ionizing radiation which would kill 50% of a large batch of organisms within a specified period of time.

Median lethal time (LT_{50}): In a time dependent biological assay procedure, this is the period of exposure to a pathogenic (including toxicological) stimulus which will produce death in half the test subjects. The length of exposure is a direct measure of dosage and an increase in the period of exposure results in an increasing uptake and true dose increases in the same ratio. Its symbol is LT_{50} (also see MEDIAN SURVIVAL TIME ST_{50}).

Median nerve cord: The median strand of nerve tissue produced from the ventral neuroblasts.

Median nerves: Unpaired nerves arising from the ganglia of the ventral nerve cord between the roots of the connectives.

Median oviduct: In insects, the single duct formed by the merging of the paired lateral oviducts; this duct opens posteriorly into a genital chamber or vagina.

Median survival time or ST_{50}: A restricted concept of median effective time, the time at which death occurs in half the test subjects, after exposure to a pathogenic (including toxicological) stimulus. ST so is not a direct measure of dosage, unlike LT_{50}.

Medio-cubital cross vein: A cross vein connecting the media and cubitus.

Medlure: A synthetic attractant for baiting the Mediterranean fruit fly; see butyl 4(or 5)-chloro-2-methyl-cyclohexanecarbosylate; product of Universal Oil.

Membrane: A thin film of tissue, usually transparent; that part of the wing surface between the veins; the thin apical part of a hemelytron (Hemiptera).

Mentum: The distal part of the labium, which bears the palpi.

Meropleuron (pl:, meropleura): A sclerite consisting of the meron (basal part) of the coxa and the lower part of the epimeron.

Meropodite: The fourth segment of a generalized limb. (Femur.)

Mesenteron: The stomach of the embryo; in insects regenerated from scattered endodermal cells or from intact endodermal remnants; becomes the epithelium of the adult stomach or ventriculus.

Mesoblast: The middle cell tissue of the embryo, including mesenchyme and mesoderm.

Mesoderm: In insects, the embryological tissue that forms a middle layer and gives rise to muscles, heart, blood cells, fat body, reproductive organ, and others.

Mesonotum: The dorsal sclerite of the mesothorax.

Mesopleuron (pl:, mesopleura): The lateral sclerite (s) of the mesothorax; the upper part of the episternum of the mesothorax (Diptera).

Mesoscutellum: The scutellum of the mesothorax.

Mesoscutum: The scutum of the mesothorax.

Mesosoma: The middle of the three major divisions of the insect body. In most insects it is the strict equivalent of the thorax, but in higher Hymenoptera it includes the propodeum.

Mesosternum: The sternum, or ventral sclerite, of the mesothorax.

Mesothorax: The second segment of the thorax; bearing the first pair of wings in winged insects.

Metabola: Insects having a definite metamorphosis.

Metabolism: The whole process or series of changes of food into tissues and cell-substances and of these latter into waste products; the first of these changes being anabolic or building up, and the second katabolic (catabolism) or breaking down.

Metabolite: A compound derived in the case of a pesticide by chemical action upon the pesticide within a living organism (plant, insect, higher animal, etc.). The action varies (oxidation, reduction, etc.) and the metabolite may be either more toxic or less toxic than before. The same derivative may in some cases develop upon exposure of the pesticide outside a living organism.

Metacarpus: The posterior sub division of the esophageal corpus; usually ovate in shape and frequently containing a valve.

Metamorphosis: 1. Changes in form during the development of an insect from the egg to the adult. 2. The series of changes through which an insect passes in its growth from the egg through the larva and pupa to the adult; complete when the pupa is inactive and does not feed; incomplete when there is no pupa or when the pupa is active and feeds. 3. The change in form of an organism through various growth stages from larva to adult, e.g. a caterpillar into a pupa and subsequently into a fully developed insect.

Metanotum: The dorsal sclerite of the metathorax.

Metapleuron (pl:, metapleura): The lateral sclerite (s) of the metathorax.

Metapneustic: An insect respiratory system (particularly in some dipterous larvae) in which only the last abdominal pair of spiracles is functional; cf. Amphipneustic.

Metascutellum: The sternum, or ventral sclerite, of the metathorax.

Metasoma: The hindmost of the three principal divisions of the insect body. In most insect groups it is the strict equivalent of the abdomen. In the higher Hymenoptera it is composed only of some of the abdominal segments, since the first segment (the "propodeum") is fused with the thorax and has therefore become part of the mesosoma.

Metasternum: Ventral sclerite of the metathorax.

Metatarsus (pl:, metatarsi): The basal segment of the tarsus.

Metathorax: The third segment of the thorax. It bears the hind legs and the second pair of wings in winged insects.

Metepimemeron (pl:, metepimera): The epimeron of the metathorax. Microbial "insecticide". 1. A pathogenic microorganism or its products (e.g. toxins) when used by man to suppress an insect population; use of the word "insecticide" is properly reserved only for chemical agents of insect mortality, and microbial pathogen is a preferable term for materials whose active agent is a microorganism. 2. Pathogen used for killing insects. Free from harmful residues, non-phytotoxic to crops of higher doses, harmless to beneficial insects.

Microenvironment: A small or restricted set of distinctive environmental conditions, such as dead animal or fallen log, (at habitat).

Microgram: A metric weight measurement equal to 1/1,000,000th of a gram; approximately 28,500,000 micrograms equal 1 ounce

Micron: 1. One thousandth part of a millimeter. 2. A unit to measure the diameter of spray droplets. Approximately 25,000 (microns) equal an inch.

Microorganism: A microscopic animal or organism.

Microphagous: Feeding on organisms of minute size.

Micropyle: A minute opening or group of openings into the insect egg through which the spermatozoa enter in fertilization.

Midgut: The middle part of the alimentary canal and the main site of digestion and absorption.

Migration: 1. A regular movement from one region to another by animals especially birds. 2. Movement of ions (and nutrients) in soil or during electrolysis. 3. In chromatography, movement of loaded sample in the direction of mobile phase during elution or development.

Mimicry: Superficial resemblances between two organisms that result in protection for at least one organism, the mimic.

Mite: 1. Tiny eight-legged animal with a body divided into two parts. It has no antennae (feelers). During the nymphal stage it has six legs. 2. A sub-class of the Arachnida allied to ticks, found in many habitats (e.g. the soil, decaying organic matter) and parasitic on both plants and animals.

Miticide: Any chemical mixture or compound, such as parathion used for killing mites.

Molar solution: A solution containing one gram mole of solute per cubic decimetre (litre) of the solution.

Mole: The unit of amount of the substance in gram molecular weight of that substance (The gram molecular weight is the molecular weight in grams of one mole of any substance as compared to one gram atom of ^{12}C taken equal to 0.012 kilogram).

Molluscicide: A compound used to control slugs and snails which are intermediate hosts of parasites of medical importance to man.

Molt (moult): The casting off of the outgrown skin or exoskeleton in the process of growth. Also the cast-off skin itself. The word is further used as an intransitive verb to designate the performance of the behaviour.

Moniliform antenna: The antenna made up of bead-like segments.

Monitoring system: A regular system of keeping track of and checking up on whether or not pesticides are escaping into the environment.

Monocondylic joint: A joint with a single point of articulation between the adjacent leg segments.

Monognatha: Forms having single jaw.

Monophagous: Feeding upon only one kind, for example one species or one genus of plants.

Moribund (insects): Dying, near death. The insects with no sign of recovery.

Mortality: Death-rate.

Motor nerves: Nerves carrying neurone fibres which carry impulses from the central nervous system (CNS) to the effector organ (ct. sensory nerves).

Motor neuron: A neuron leading from the central nervous system to an effector organ.

Moulting: The process of casting-off the old tough outer layer (cuticle) of the body wall to accommodate the growth of the body.

Moulting hormone: The hormone responsible for moulting in insects.

Mouth: Strictly used for the opening through which food enters the alimentary canal.

Mouthparts: Collective name for labrum, mandibles, maxillae, labium and appendages.

Multi-dose poison: Chronic poisons, such as anticoagulants like WARFARIN.

Multiparasitism: A condition resulting from the simultaneous use of a single host individual by two or more species of primary parasitoids; ct. superparasitism.

Multivoltine: Having two or more complete generations annually: ct. univol tine.

Muscle: Fleshy fibers that serve tomove the appendages and other organs of the body.

Mutagenic: Capable of producing genetic change.

Mutant: An individual having an abnormality of structure, properties or behaviour in which it differs distinctly from the type and which has internal (genetic) rather than environmental origin, so that there is probability of transmission to off spring.

Mutation: 1. Any relatively stable heritable change in the genetic material. 2. Process that results in gene changes from one heritable allele to another heritable allele.

Mutualism: Symbiosis that benefits the members of both of the participating species.

Myoneural junction: The site of transmission of an impulse from a nerve to a muscle.

Natality: Birth-rate.

Natural control: 1. The reduction of pest populations by the forces of nature, such as climatic factors, parasitoids, predators and diseases. 2. The process of dynamic equilibrium which maintains the characteristic mean density of a wild population within particular upper and lower limits, over a period of time, by a complex combination of all the additive, conditioning and subtractive processes impinging upon that wild population.

Natural enemies: 1. Strictly the parasitoids, predators, and pathogenic microorganisms associated naturally with a specific wild population of plants or animals, and causing mortality or debility to the individuals thereof; often used in a general sense for all parasitoids, predators, and pathogens. 2. The predators and parasites in the environment which attack pest species.

Natural selection: 1. Differential survival and reproduction of individuals. 2. An increase in the frequency of occurrence of some genes or gene combinations and a decrease in the frequency of others because of the selective action of the environment.

Necrosis: 1. Death of part of a living body. 2. Localized death of living tissue, i.e., death of a certain area of a leaf or of a certain area of an organ.

Nematicide: A material, often a soil fumigant used to control nematodes infesting roots of crop plants.

Nephrocytes: Special cells scattered in groups in various parts of the body, but especially abundant in the dorsal sinus, where they are known as pericardial cells; probably have some function in connection with excretion and have been termed "hepatic cells."

Nerve: Anyone of the fibrous tracts of the peripheral nervous system, whether a single fiber or a group of fibers; a nerve trunk.

Nerve cord: Cord composed of delicate filaments of tissue functioning to transmit stimulus to or from a ganglion or from or to any part of the body.

Nerve fiber: The axon or other branches of a neurocyte.

Nerve tract: A strand of nerve fibers; usually applied to tracts within a nerve center.

Nerve trunk: A bundle of nerve fibers in the peripheral system; the usual nerves.

Nervous system: All nerve cells and tissues in animals, including the brain, spinal cord, ganglia, nerves, and nerve centers.

Neural groove: The median ventral groove of the embryo between the neural ridges.

Neural ridges: The two longitudinal ventral ridges of the embryo in which are formed the lateral cords of neuroblasts.

Neurilemma: The nucleated sheath or nerve tissue, covering the ganglia, nerve trunks, and terminal branches.

Neuroblasts: The primitive nerve cells differentiated from the ectoderm.

Neurone: The entire nerve cell including all its processes.

Neuropile: The medullary substance, or mass of fibrous tissue within a ganglion.

Neurosecretory: Pertaining to the secretion of hormones by nerve cells.

Niche: The precise constellation of environmental factors into which a species fits or which is required by a species.

Nitrification: The conversion of the atmospheric nitrogen into nitrates by bacteria.

No-residue: As the term applies to pesticides the act of registration of an economic poison on the basis of the absence of a residue at time of harvest on the raw agricultural product when the economic poison is used as directed.

Nocturnal: 1. Pertaining to the night hours. 2. Active by night.

Nomadic phase: The period in the activity cycle of an army ant colony during which the colony forages more actively for food and moves frequently from one bivouac site to another. At this time the queen does not lay eggs and the bulk of the brood is in the larval stage opposed to statary phase.

Nonoccluded viruses: Viruses which do not occur within inclusion bodies; also called noninclusion virus.

Nonpolar insecticides: The molecules of nonpolar insecticides, such as DDT, BHC, and synthetic pyrethroids, do not have any separation of charges, because the electrons are uniformly distributed such insecticides are lipophilic, i.e. have affinity for fat or fat-like substances.

Nonpreference: 1. An insect pests response to a potential host (variety) which displays various characteristics that discourage its use for oviposition, food or shelter. 2. In this case, the resistance of a plant to the attack of a pest is due to the lack of one or more of the preferred factors in the plant.

Nonrenewable resources: A natural, normally nonliving, resource such as a mineral which is present in finite supply and is not renewed by natural system.

Notopleural bristles: Bristles on the notopleuron (Diptera).

Notopleuron (pl:, notopleura): An area on the thoracic dorsum, at the lateral end of the transverse suture (Diptera).

Notum: 1. The dorsal part of segment; usually refers to those of the thorax. The prefix pro, meso, or meta, indicates which segment. 2. The dorsal surface of a body segment (usually used when speaking of the thoracic segment).

Nozzle: 1. Mouthpiece of a spout through which liquids are. discharged in a stream or as a spray. 2. Devices which control drop size, rate, uniformity, thoroughness, and safety of a pesticide application. The nozzle type determines the ground pattern and safety of a pesticide application. Examples: Flat fan, even. flat fan, cone, flooding, off-set, atomizing, broadcast, and solid stream nozzles.

Nozzle: functions of various types of nozzles used in pesticide spray operations have three basic functions i.e. (i) to break up the liquid spray formulation into suitable droplets; (ii) to direct the droplets on target; and (iii) to check the quantity of the spray liquid dispersed.

Nozzle types: 1. Air blast nozzle using high velocity air to break up the spray liquid supplied at low pressure. 2. Anvil Nozzle in which the spray liquid jet strikes a smooth, solid surface at a high angle of incidence. 3. Cone (or swirl) nozzle in which the liquid emerges from the orifice with tangential velocity imparted by passage through one or more tangential or helical channels in the swirl chamber. 4. Hollow cone nozzle in which spray jet has a core of air breaking to give drops in an annular pattern. 5. Fan nozzle the aperture is an elongate horizontal slit, producing a fan-shaped spray pattern. 6. Deflector nozzle in which a fan-shaped sheet of spray is formed by directing the liquid over a sharply inwardly curving surface.

Nucleus: The spheroid body within a cell that has the major role in controlling and regulating the cell's activities and contains the hereditary units or genes.

Nuptial flight: The mating flight of the winged queens and males.

Nurse cells: Cells that are located in the ovarian tubes of certain insects and that furnish nutriment to the developing eggs.

Nymph (pl: nymphae): 1. Immature stage of an insect species, similar to adult but with wings and sex organs not fully developed. 2. In general entomology, the young stage of any insect species with hemimetabolous development. In termites, the term is used in a slightly more restricted sense to designate immature individuals which possess external wing buds and enlarged gonads and which are capable of developing into functional reproductives by further molting. 3. In insects, an immature stage (instar) which is similar to the adult except for absence of functional wings and sexual organs. Distinct form larva of other insects, which is entirely different from the adult.

Obligate parasitism: A parasite which is obliged to follow a parasitic mode of the life and can not exit in any other way.

Obligate pathogen: A disease causing microorganism which requires a living host in which to grow and reproduce.

Oblique vein: A slanting cross vein in Odonata where *Rs* crossed M_{1+2}.

Obtect: A pupa with legs and wingsfirmly glued to the body e.g. Lepidoptera.

Occipital arch: The area of the cranium between the occipital and postoccipital sutures; its dorsal part is the occiput proper (Oc), its lateral parts the postgenae (Pge).

Occipital condyles: Processes on the margin of the postocciput to which the lateral neck plates are articulated.

Occipital ganglion: A single or paired postcerebral ganglion of the stomodaeal nervous system. (Pharyngeal, oesophageal, or hypocerebral, ganglion.)

Occipital suture: A transverse suture in the posterior part of the head which separates the vertex from the occiput dorsally and the genae from the postgenae laterally.

Occiput: Portion of head behind the vertex, eyes, and gena.

Occluded viruses: Viruses characterized by enclosure of the mature replicated virions in the matrix of proteinaceous or crystal-like inclusion bodies, such as capsules or polyhedra; also called INCLUSION VIRUSES.

Occlusor muscle of a spiracle: A muscle serving to close either an outer or an inner closing apparatus of the spiracular atrium.

Ocellus: One of the three simple eyes of adult insects, located on or near the center line of the dorsal surface of the head. The ocelli should be distinguished from the laterally placed compound eyes.

Ocular sclerite: A narrow band of the cranial wall encircling the compound eye within the ocular suture.

Ocular suture: The line of inflection in cranial wall around the compound eye, forming internally a circumocular ridge.

Oenocytes: Large ectodermal cells occurring in the abdomen of most insects, sometimes within the epidermis, but generally free in lateral segmental groups, or scattered through the fat body. (Cerodecytes).

Oesophagus: A tubular part of the fore-gut between the mouth and the crop.

Olfactory organs: The organs of smell.

Oligophagous insects: Insects, such as the spotted boll-worm, which restrict their feeding to a small number of usually similar plants.

Oligophagus parasite: A parasite which can develop upon a few of the closely related host species.

Oligopod: Larva coming out from embryo hatched in advanced stage of development, bearing a general resemblance to adult e.g. Thysanoptera.

Ommatidium: One of the basic visual units of the insect compound eye. The ommatidia are bounded externally by the facets that together make up the glassy, rounded outer surface of the eye.

Omnivorous: Feeding generally on animal or vegetable food, or on both.

Onisciform larva: A flattened platyform larvae like a woodlouse in appearance.

Ontogeny: The development of the individual as distinguished from that of the species.

Oocytes: The egg cell differentiated from the oogonium, before maturation.

Oogonium: The first stage in the differentiation of an egg cell from a primary female germ cell.

Ootheca (pl:, oothecae): An egg case formed by the secretions of accessory genital glands or oviducts, as in cockroaches.

Operculum (pl:, opercula): A lid or cover.

Opisthognathus: Having the front of the head deflected backward so that mouth parts project to the rear.

Optic lobes: Lateral expansions of the brain of an insect innervating the compound eyes and ocelli.

Oral toxicity: 1. Toxicity of a compound when given by mouth. Usually expressed as the number of mg of chemical per kg of body weight of animal when given orally in single dose that kills 50 percent of the animals. The smaller the number, the greater the toxicity. 2. Ability of a pesticide chemical to sicken or kill an animal or human if eaten or swallowed.

Orbital plate: A sclerite on the head next to the compound eye (Diptera); also called genovertical plate.

Organelles: Organized microstructures inside cells to which certain biochemical processes are confined e.g. mitochondria, chloroplasts.

Organism: Any living thing; plant animal, fungus, bacteria, insect, etc.

Orthene (acephate): Low toxicity. Orthene is particularly useful for controlling aphids, grasshoppers, loopers, thrips, and caterpillars on flowers and other types of plants.

Osmoregulation: Regulation of the salt and water concentration in cells and organisms.

Osmosis: The transfer of materials that takes place through a semipermeable membrane that separates two solutions, or between a solvent and a solution that tends to equalize their concentrations. The walls of living cells are semipermeable membranes.

Ostia (sing: ostium): Openings on the sides of the heart through which the blood enters.

Ostial valves: Valve-like pouches of the heart wall containing the ostia at their inner ends.

Ovariole: One of the egg tubes which collectively form the ovary.

Ovary: A mass of tubes lying one on each side of the body cavity of the female insect in which the eggs are developed.

Ovicide: 1. A chemical which kills eggs before they hatch or one so used. Almost restricted to action on eggs of phytophagous mites (red spider). The word is almost universally accepted in this sense, but the classical scholar insists that it should mean "sheep (Latin, ovis) killer" and that an egg-killer should be "ovacide". 2. Pesticides oragents used to destroy insect, mite or nematode eggs.

Oviduct: One of the paired tubes through which the egg passes from the ovary into the vagina.

Oviparae: In aphids, the oviparous lemales.

Oviparous reproduction: Reproduction by means of eggs, which hatch outside the body.

Oviporus: The posterior opening of the vagina in most Lepidoptera, serving only for the discharge of the eggs when there are two genital apertures.

Oviposition: The act 01 laying or depositing eggs.

Ovipositor: The extensible tube on the abdomen of a female insect which she uses to place her eggs in suitable location. Often capable of remarkable drilling power. Hence oviposition the act of egg-placing.

Ovoparity: Reproduction by laying eggs.

Ovoviviparity: Reproduction in which eggs hatch soon alter they are laid.

Palaeoentomology: Study of extinct fossil insects.

Palp: Sensory feelers in the mouth parts of insect. There are usually two pairs of palps, one pair (maxillary palps) attached to the maxilla, and the other pair (labial palps) attached to the labium.

Palpifer: Small lobe or sclerite articulated to the stipes which bears the maxillary palpus.

Palpiger: Small sclerite on the labium which bears the palpus.

Paraglossa (Pl: paraglossae): One of a pair of lobes at the apex of the labium. Laterad to the glossae.

Paranotal lobes: Lateral lobes of the pronotum in certain fossil insects, and theoretical lobes of the mesonotum and metanotum supposed to be the precursors of the wings.

Paranotum (pl: paranota): Notal flanges in fossil insects.

Paraproct: One of a pair of lobes bordering the anus lateroventrally.

Parapsidal furrows: Lateral grooves divergent anteriorly in the posterior part of the scutum of the mesothorax of some Hymenoptera.

Parapsidal suture: A longitudinal suture of the mesonotum spearing the median area from the lateral area (Hymenoptera).

Parapsides: Lateral areas of the mesoscutum in some Hymenoptera laterad of the parapsidal furrows. (Singular, parapsis.)

Parasite: An organism which lives and feeds in or on another. Ectoparasites lives on the surface of the host, Endoparasites within its body. Facultative parasitic can live either parasitically or free, obligate parasitic live only parasitically.

Parasitism: The state of being infected with parasite.

Parasitoid: The insects which parasitise other insects.

Parasporal body: An inclusion that develops with the endospore in the cells of certain bacteria during the process of sporulation. In *Bacillus thuringiensis* and other crystalliferous bacteria, the parasporal body is in the form of a toxic crystal.

Parietals: The lateral areas of the cranium between the frontal and occipital areas, separated above by the coronal suture

Parthenogenesis: Reproduction by the development of an egg without its being fertilized by a sperm.

Partition coefficient: It is the ratio of concentrations of solute in two phases after partition between two immiscible phases i.e. $Kd=Cs/Cm$, where Cs and Cm are concentrations of the solute in the phases.

Parts per million: Proportion of toxicant present, in relation to that of plant material on which it has been deposited. Usually applied in connection with the edible portion of a crop and its suitability for consumption.

Parts per billion: A way of expressing amounts of chemicals in foods, plants, animals, etc. One part per billion equals 1 kg. in 1000,000 tons.

Parts per million: A way of expressing amounts of chemicals in foods, plants, animals, etc. One part per million equals 1 kg. in 1000 tons.

Pectines: A comblike or rakelike structure (Hymenoptera).

Pectinate: Comb-like; with branches or processes like the teeth of a comb, as in pectinate antennae.

Pectinate antennae: Antennae possessing one row of projections like the teeth of a comb; comb-like antennae.

Pedicel: The "waist" of the ant. It is made up of either one segment (the petiole) or two segments (the petiole plus the postpetiole).

Pedicel: 1. One of the ovariole stalks (*Eirohrenstielen*), or short ducts from the egg tubes to the oviduct. 2. The second segment of the antenna; the stem of the abdomen, between the thorax and the gaster in ants.

Pellet: 1. A dry formulation of pesticide mixed with other components in discrete particles, usually larger than 10 cubic millimeters. 2. Seed coated with inert material, often incorporating pesticides, to ensure uniform size and shape for precision drilling.

Penetration: The ability to get through; the process of entering.

Penis: One of the paired intromittent organs of certain insects, or the usual median phallic organ.

Pericardial cells: Strands of special cells along the sides of the heart, probably in most cases nephrocytes.

Periodic release: A method of beneficial organism introduction which involves repeated liberations to artificially maintain high population levels of indigenous biotic agents in situations where such levels are unattainable naturally.

Periodicity: A regular cyclic or repetitive behaviour of an organ cell or 'organism in a unit time.

Peripheral nervous system: The outlying parts of the nervous system in distinction to the central ganglia and connectives, including the sensory neurocytes and their axons, and the axons of the motor neurones.

Peripneustic: With none or only a few spiracles closed in each lateral series.

Peripneustic respiration: The prevalent type in insects, consisting of a row of spiracles on each side of the body.

Periproct: The terminal piece of the body containing the anus, anterior to which segmentation takes place.

Peristome: The ventral marginal part of the cranium formed by the clypeus, or epistoma, the pleurosomata, and the hypostomata.

Peritreme: A sclerite of the body wall containing the spiracular opening.

Peritrophic membrane: 1. The delicate membrane surrounding the food in the mid-gut; present in many insects. 2. A cylindrical membranous envelope surrounding the food in the ventriculus, and sometimes extending into the proctodaeum; generated from the ventricular epithelium, either from all or the length of the latter or from a ring of specialized cells at its anterior end.

Perivisceral sinus: The principal part of the body cavity, between the dorsal diaphragm and the ventral diaphragm if the latter is present.

Permissible level: A concentration expressed in ppm of residue in or on a food when first offered for consumption. It is calculated from (i) the Accepted Daily Intake (ADI), (ii) the food factor, and (iii) the average weight of the consumer. The ADI is the daily dose which during an entire life time will not cause any injury. It is set at a safe level as determined by data obtained in animal feeding studies. A safety of 100 is used in translating animal data to human beings. It is expressed in milligram of the pesticide per kilogram of body weight per day (mg/kg/day). The food factor refers to the total diet made up of a food or of various foods.

Persistance: 1. The term applied to chemicals that remain active for a long period of time after application. 2. The quality of an insecticide to persist as an effective residue due to its volatility and chemical stability. 3. The property of a chemical to remain effective as a residue. 4. Property of chemical to remain active for long period after application.

Pest management: 1. A philosophical component of overall resource management involving the decision making process that considers the value and impact of pest suppression in relation to all other aspects of human societal endeavor and environmental integrity. 2. Many workers consider pest management and integrated control as synonymous. However, a distinction between the two strategies may be made on the basis of number of control techniques involved in them. The pest management may involve the use of even a single control method, whereas the integrated control invariably makes use of two or more methods. Anyway, both the control approaches attempt to avoid harmful side-effects on the agro-ecosystem, and the pest control is based on sound ecological and economic principles.

Pest: 1. It is an organism whose activities cause it to be inimical to the welfare of human beings. It includes invertebrates such as insects, mites and ticks and some vertebrates, such as birds and rodents. 2. Troublesome or destructive insect, animal, plant or other organism; causes injury or loss to growing crop or injury or irritation to domestic animal or man. 3. Organism that causes loss or annoyance to man or his property to the extend that warrants its control. 4. Animals, insects, fungi, weed, etc., which are troublesome to crops or livestock, or cause harmful diseases, and result in considerable losses in agricultural or horticultural production in terms of both quantity and quality. 5. By law, forms of plant and animal life and viruses when they exist under circumstances that make them injurious to plants, man, domestic articles or substances.

Pest control: The use of disinfectant, herbicide, pesticide, insecticide, management or cultural practice to control pests.

Pest control programme: An organised programme to keep down the pest population in an area below pest density.

Pest density: Popularion level at which a pest species causes economic damage.

Pest spectrum: The complete range of pests attacking a particular crop.

Pesticide residue: Pesticide remaining on or in a plant, plant product air, water, or treated area following a time lapse after application.

Pesticide tolerance: The amount of pesticide residue which may legally remain in or on a food crop. Federal residue tolerances are established by the Environmental Protection Agency.

Pesticides, botanical: Plant products used for pest control; include insecticides (pyrethrins, rotenone, nicotine, neem oil), synergists (dillapiole, sesamin, sesamolin), antifeedants (azadirachtin, schkuhrin I and II), insect juvenile hormones (paper factor), repellents (geraniol, carvacrol), ovipositional repellents (eugenol, 1,8. cineol) etc.

pH: Symbol denoting relative concentration of hydrogen ions in solution; a scale of relative acidity and alkalinity.

Phallobase: The proximal part of the phallus, highly variable in its development, sometimes a large structure supporting the aedeagus, often produced into a thecal fold or sheath about the aedeagus sometimes represented only by basal phallic sclerites in the wall of the genital chamber.

Phallus: The unpaired penis, or median intromittent organ, including the phallobase, the aedeagus, the endophallus, and various processes of the phallobase and the aedeagus if present.

Pharynx: 1. The back part of the mouth; a slight enlargement of the beginning of the oesophagus. 2. The part of the stomodaeum between the mouth or buccal cavity and the oesophagus, the dorsal dilator muscles of which arise on the frons and the dorsal part of the cranium and are inserted posterior to the frontal ganglion and its connectives; usually not extending beyond the nerve ring of the head, but in some insects there is a posterior pharynx behind the brain.

Phenotype: The physical manifestation of a genetic trait.

Pheromone: 1. A chemical substance, usually a glandular secretion, which is used in communication within a species. One individual releases the material as a signal and another responds after tasting or smelling it. 2. A special chemical substance which, secreted by one individual, influences the behavior of other individuals of the same species. 3. Chemicals produced by insects and other animals to communicate with and influence the behavior of other animals of the same species. 4. A substance secreted by an insect to the exterior causing a specific reaction in the receiving insects. 5. An exocrine chemical messenger secretion produced by an organism, which influences or activates the behavior or physiology of other individuals of the same species. 6. A substance secreted by an insect for giving information to other insects of the same species and, thereby, influencing their behaviour.

Photoperiodism: 1. A response by an organism to the duration or pattern of light and dark periods. 2. Seasonal responses to changes in length of daylight.

Photoreceptor: Sense responsive to light. Phragma (pl. phragmata). A platelike apodeme of the dorsal wall of the thorax.

Phragmata: Plate-like apodemal lobes of the antecostae of the mesonotum, metanotum, and 'first abdominal tergum; the second and third carried by the postnatal plates of the pterothorax when the latter are separated from the following terga to which they normally organ belong. (Singular phragma).

Phylogeny: Evolutionary history of species.

Physiology: Study of various functions of living beings.

Phytophagous: Feeding on plants or on materials of plant origin.

Phytoplankton: Microscopic plant life living suspended in water.

Phytosanitation: Measures requiring the removal or destruction of infected or infested plant material likely to form a source of reinfection or reinfestation.

Phytotoxic: Toxic to plant, usually restricted to higher (green) plants. Hence phytotoxicity, which is used generally to include the action of intentional herbicides but by some authors restricted to adverse side effects on the crop of fungicides, insecticides or formulating agents.

Plant resistance: The ability of plant to grow and produce economically despite the presence of a pest. Such plants possess some chemicals or juices which inhibit the parasite from feeding on it.

Plastron: 1. A film of air over part of the body of some aquatic insects; like a bubble, it serves as a physical gill. 2. A bubble of air trapped in the curved hydrofuge hair on the body of an aquatic insect. It is used for respiration. 3. Structure to obtain dissolved oxygen.

Platyform larva: A larva that is extremely flattened, as the larva of the Psephenidae.

Pleura (sing: pleuron or pleurum): The lateral areas of the thoracic segment.

Pleural: Pertaining to the pleural, or lateral sclerites of the body; lateral.

Pleural apophysis: The internal arm of the pleural ridge.

Pleural region: The podial region, or ventrolateral parts of the body on which the limbs are implanted, metamerically divided into segmental pleural areas.

Pleural ridge: The endopleural ridge formed by the pleural suture, bracing the pleuron above the leg, or bet,veen the coxal articulation and the wing support.

Pleural sulcus: Groove running dorsally from coxa; often referred to as pleural suture.

Pleural suture: The external groove at the pleural ridge, separating the episternum from the epimeron.

Pleural wing process: The wing support of the pleuron at the upper end of the pleural ridge.

Pleurite: Any minor sclerite of the pleural area of a segment, or one of the component sclerites of a pleuron.

Pleura ventral line: The line of separation between the pleural region and the venter; lying mesad of the limb bases, but obscured when the latter are fused with the sterna.

Plica jugalis: The jugal fold of the wing of some insects, or radial line of folding setting off the jugal region from the vannal region. (Axillary furrow, plica ano-jugalis).

Plica vannalis: The vannal fold of the wing, or radial line of folding usually between the cubital field and the first vannal vein, but somewhat variable in position (Anal furrow plica analis).

Plumose antennae: Feather-like antennae.

Point of drip or runoff: When a spray is applied until it starts to run or drip off the ends of the leaves and down the stems of plants or off the hair or feathers of animals.

Poison bait: 1. A poison mixed with bran, molasses, grain, or other attractant so as to serve as a bait and kill insect pests and rodents. 2. It consists of a relatively small quantity of a poison combined with a food material of the pest. 3. An attractant foodstuff for insects, molluscs, or rodents, mixed with an appropriate toxicant.

Polar chemical: A chemical compound composed of molecules with a dipole (a pair of equal and opposite charges separated by a small distance), such as water, alcohol, and salt solutions.

Pollen brush: A brush of short stiff hairs used in collecting pollen (bees).

Pollen stores: Bumblebee species that store pollen in abandoned cocoons. From time to time the adult females remove the pollen from the cocoons and feed it into a larval cell in the form of a liquid mixture of pollen and honey (op-posed to pouch makers).

Pollutant: A harmful chemical or waste material discharged into the water, soil, or atmosphere. An agent that makes something dirty or impure.

Polyembryony: The development of several individual organisms from a single egg.

Polygamy: The condition of a single male having many female mates.

Polygenic: A characteristic, such as resistance, conferred by more than one gene.

Polyhedral viruses: Viruses which cause insect disease characterized by the presence of polyhedral (many sided) inclusions in the infected cells of the host.

Polyphagous insects: Insects, such as locusts and grasshoppers, which feed on a variety of plants.

Population dynamics: The study of numerical changes in populations of living organisms in time and space and of the processes which cause such variations.

Pore canal: The channel of the cuticula beneath the seta or other external part of many sense organs.

Posterior cross vein: A cross vein at the apex of the discal cell (Diptera).

Posterior intestine or rectum: The terminal section of the proctodaeum commonly termed the "rectum," but usually divided into an anterior rectal sac (rsc), and a posterior rectum proper (rect). (Mastdarm).

Posterior notal wing process: A posterior lobe of the lateral margin of the alinotum supporting the third axillary sclerite of the wing base. (Hinterer Gelenkfortsatz).

Postfrontal sutures: Facial sutures present in some insects diverging from the coronal suture laterad of the antennal bases.

Post gena: That part of the head which extends downward and back of the gena.

Postlabium. The basal or proximal part of the labium.

Postmarginal vein: The vein along the anterior margin of the front wing beyond the point where the stigma vein arises (Chalcidoidea).

Post mentum: The basal portion of the labium proximad of the labial suture.

Postnodal cross vein: A series of cross veins just behind the costal margin of the wing, disted of the nodus, and extending from the costal margin of the wing to R_1 (Odonata).

Postnotum: An intersegmental plate of the dorsum of the thorax associated with the tergum of the preceding segment, bearing the antecosta and usually a pair of phragmatal lobes. (Ph ragmanotum).

Postnotum phragmanotum: The postscutellar, phragma-bearing plate often present in the dorsum of an alate segment, derived from the anterior part of the following tergum. (Postscutcllum).

Postoccipital suture: The posterior submarginal groove of the cranium having the posterior tentorial pits in its lower ends; internally it forms the postoccipital ridge (PoR) on which are attached the dorsal prothoracic and neck muscles of the head.

Postocciput: The extreme posterior rim of the cranium behind the postoccipital suture, probably a sclerotic remnant of the labial somite.

Postscutellum: A small transverse piece of a thoracic notum immediately behind the scutellum; in Diptera, an area immediately behind or below the mesoscutellum.

Poststernite: The postcostallip of a definitive sternal plate that includes the intersegmental sclerotization.

Posttergite: The narrow postcostallip of a postnotal thoracic plate.

Potency: The strength of something. Example: how deadly a poison is.

Potentiation: The joint action of two pesticides to bring about an effect greater than the sum of their individual effects.

Pre-oviposition period: The period of time between the emergence of an adult female insect and the start of its egg laying.

Precosta: A small first vein of the wing in certain fossil insects.

Precoxal bridge, precoxale: The precoxal part of the pleuron anterior to the trochantin usually continuous with the episternum, frequently united with the sternum, sometimes a distinct sclerite.

Predator: 1. An animal (or, rarely, a plant) which feeds upon other animals, called its prey. The prey is killed and consumed by its predator which generally devours many individuals during its life (contrast parasite). There are borderline cases between predation and parasitism, e.g. species of wasp which inject a permanent paralyzing drug into a caterpillar before laying eggs in it. 2. An insect or other animal that attacks, feeds on and destroys other insects or animals. Predators help to reduce the number of pests which cause disease, damage and destruction. 3. Any animal, including insects, that preys upon other animals and devours them. Beneficial when they feed upon plant parasites. 4. It is free living organism that kills its prey.The predator is usually bigger than its prey and requires more than one prey to complete its development. 5. An animal which feeds upon other animals (prey) that are usually smaller and weaker than itself, frequently devouring them completely and rapidly. A predator most often is required to seek out and attack more than one prey to reach maturity.

Pregenital: Anterior to the genital segments of the abdomen.

Prelabium: The distal part of the labium comprised of the prementum, the ligula, and the palpi.

Prementum: The distal part of the labium, distad of the labial suture, on which all the labial muscles have their insertions.

Preoral: Anterior to or in front of the mouth.

Prescutal suture: A transverse groove of the mesonotum or metanotum behind the antecostal suture, setting off a prescutum from the scutum, and forming internally a prescutal ridge (TR).

Prescutum: The anterior area of the mesonotum or metanotum between the ante costal suture and the prescutal suture, when the latter is present.

Presternum: A narrow anterior area of the sternum sometimes set off from the basisternum by a submarginal suture of the eusternum. (Not the acrosternite).

Pretarsus: The terminal parts of the leg distal to the tarsus, including median remnants of the dectylopodite, and the lateral claws, or ungues; in most larvae a simple clawlike segment.

Preventive application: Application of control measures in anticipation of an attack of an organism or outbreak of a disease.

Prey: An animal when fed upon by a predator. A predator of one species can itself be prcy to another.

Primary parasite: A parasite which establishes itself in or upon a free living host, that is, the host is not a parasite.

Primary parasitoid: An insect parasite of any arthropod which is not itself parasitic.

Proboscis: 1. The lengthened mouthparts of an insect specially adapted for piercing, sucking or other specialized modes of feeding. 2. The extended beaklike mouth parts of sucking insects.

Procephalon: The region of the definitive head anterior to that of the gnathal segments, representing the primitive protocephalon, formed in the embryo from the cephalic lobes and the tritocerebral segment; bearing the labrum, the mouth, the eyes, the first antennae, and the second antennae when the last are present.

Proctodaeum: The hind intestine, including the Malpighian tubules.

Proctodaeum: The posterior ectodermal part of the alimentary canal. (Hind gut, hind intestine, Hinterdarm, Enddarm).

Proctodeal valve: In insects, a valve in the anterior end of the hindgut that serves as an occlusor mechanism.

Proctodeum: 1. The hindgut and Malpighian tubules of an insect. 2. Posterior division of alimentary canal formed from invagination.

Prognathous: Having the head horizontal with the jaws directed forward.

Proleg: A fleshy abdominal leg found in caterpillars (Lepidoptera, sawtlies), bearing a characteristic arrangement of crochets.

Pronotum: A large, rounded, shield like plate covering the front part of the body; the dorsal cuticular covering of the first thoracic segment.

Prophylactic: Adj. used to qualify a chemical or treatment used to prevent a pest or disease producing organism from invading the individual or crop so treated, as distinct from one used "curatively" on the diseased individual or crop. "Eradicant" is often used in the sense of "curative" when describing the treatment of fungus diseases of crops. "Protectant" is sometimes used in the sense of prophylactic.

Propneustic: That type of respiration in which only the first pair of spiracles is functional.

Propodeum: In higher Hymenoptera, the first abdominal segment when it is fused with the alitrunk. Same as epinotum.

Prosternum: The sternum, or ventral sclerite, of the prothorax.

Protective clothing: Clothing to protect the spray operator from the toxic effects of crop protection chemicals. This may include rubber gloves, boots, apron, respirator, face mask. etc.

Prothoracic gland: One of a pair of endocrine glands located in the prothorax near the prothoracic spiracles.

Prothorax: The first segment of thorax. It bears the anterior legs, but no wings.

Protocerebrum: The first part of the arthropod brain, containing the ocular and other association centers lying anterior or dorsal to the antennal (deutocerebral) centers.

Protonymph: The second instar of a mite.

Protopamchloride: An antidote for organophosphate poisoning, but not for carbamate poisoning.

Protopod: Primary larva, little more than prematurely hatched embryo, devoid of abdominal segmentation and having rudimentary cephalic and thoracic appendages e.g. parasitic Hymenoptera.

Protopodite: The basal stalk of some crustacean limbs composed of the united coxopodite and basipodite.

Proventriculus: A specialized part of the stomodaeum immediately anterior to the ventriculus.(Gizzard, gesier, chewing stomach, Kaumagen).

Pseudopods: False legs on the abdomen of the larva of butterflies, moths and sawflies.

Pseudopupa: A coarctate larva; alarva in a quiescent pupalike condition one or two instars before the true pupal stage (Coleoptera, Meloidae).

Pseudotrachea (pl:pscudotrachae): A structure having the appearance of a trachea; a false trachea.

Pteralia: The articular sclerites of the wing base, including the humeral plate and the axillary plate or axillary sclerites.

Pterothorax: The wing-bearing portion of the thorax 01 winged insects, i.e. the meso and metathorax.

Ptilinum: 1. An eversible sac of the frons in dipterous pupae used for rupturing the puparium. 2. Eversible sac in some dipterous adults that is used to rupture the puparium. 3. In Diptera Cyclorrhapha, an organ capable of being inflated and thrust out through the frontal suture, just above the bases of the antennae, at emergence from the pupa.

Pulsatile organs: Accessory organs that are sometimes present at the bases of the antennae and the wings for pumping blood into them.

Pupa: 1. A resting stage in insect metamorphosis between larva and young adult, during which movement and feeding ceases and a great change in form occurs. 2. The pupa is the resting stage of many insects. The stage between the larva (caterpillar or maggot, etc). Some caterpillars spin a silk cocoon before they change to a pupa inside. 3. The third or chrysalis stage of insect life; insect enclosed in a case; stage in metamorphosis preceding imago. 4. The stage between the larva and the adult in insects with complete metamorphosis, a non-feeding, usually inactive stage. 5. Quiescent stage between larva and adult in holometabolius insects. Obtect-legs and wings firmly glued to body (Lepidoptera) Exarate- legs and wings free from any secondary attachment to the body (Hymenoptera) Coarctate- enclosed in hardened shell formed from last larval skin, betrays no outward indication of developing insect within (Cyclorrhpha).

Puparium: 1. The tough encasement in which a larva changes to a pupa before developing into an adult insect. 2. The thickened, hardened last larval skin within which some flies pass the pupal stage.

Pygidium: The dorsal surface of the last abdominal segment.

Pyloric: Pertaining to the passage from stomach to intestine.

Pyloric valve: A valvular fold usually situated in the pyloric region of the proctodaeum, but sometimes formed by the posterior end of the ventriculus

Quarantine: 1. Isolation of an animal because of contagious disease may be enforced. 2. The place of isolation. 3. Prohibition to prevent the introduction or spread of any dangerous insect pest or plant disease. 4. All operations associated with the prevention of importation of unwanted organisms into a territory, or their exportation from it. 5. Regulation forbidding sale or shipment of plants or plant parts, usually to prevent disease, insect, nematode, or weed invasion of an area.

Quiescent: State of relative biological inactivity induced by physical factors such as low temperature; inactive, resting, non-feeding or dormant.

Radial cell: A cell bordered anteriorly by a branch of the radius; the margin cell (Hymenoptera).

Radial cross vein: A cross vein connecting R_1 and the branch of radius immediately behind it.

Radial sector: The posterior of the two main braches of the radius.

Radioisotopes: One of the broad class of elements capable of becoming radioactive and giving off atomic energy. Some radioisotopes occur naturally, others are produced artificially. The word is synonymous with radioactive elements and includes tracer elements.

Radius: The third vein of the wing; its base flexibly attached to the second axillary.

Raptorial legs: Legs adapted for catching prey.

Receptor: A so called sense organ, or specialized structure of the integument responsive to external stimuli.

Recessive characteristic: A characteristic governed by an allele that does not express itself in the presence of another allele of the same gene. Recessive traits are only expressed in homozygous individuals.

Rectal glands: Padlike or papilliform structures on the inner wall of the rectal sac (not demonstrated to be glands).

Rectal gills: The gills within the rectum of nymphs in anisopteran dragonflies.

Rectal sac: The enlarged anterior part of the rectum, sometimes produced into a large rectal caecum.

Rectal valve: A circular or lobate fold of the proctodaeal wall between the anterior instestine and the rectum.

Rectum: In insects the posterior expanded part of the hindgut, typically pear-shaped.

Recurrent nerve: The median stomodaeal nerve extending posteriorly from the frontal ganglion.

Recurrent vein: One of the two transverse veins immediately posterior to the cubital vein (Hymenoptera).

Reflex: A functional unit of the nervous system involving the entire pathway from a receptor cell to an effector.

Relative humidity: The amount of moisture in the air compared to the total amount that the air could hold at that temperature.

Remote sensing: A method for determining the characteristics of an object, organism or community from longer distances.

Repellent: 1. A compound that is annoying to a certain animal or other organism, causing it to avoid the area in which it is placed. 2. Substance which repels insects and keeps the treated object free from their attack; coaltar, creosote, oil of citreonena, examples. 3. A chemical which has the property of inducing avoidance by a particular pest. 4. A chemical that provides some protection from an insect pest by making the potential host unpalatable or offensive to the pest. Repellents exist for the protection of people, livestock, and food and fiber crops. Bordeaux mixture (6:10:100 mixture of copper sulfate, hydrated lime, and water), applied to foliage, is one of the oldest and still most successful repellents of leafhoppers, Ilea beetles, and the potato psyllid. Other, newer, repellents exist, but generally they are not widely used as substitutes for insecticides. Some plants produce their own repellents, accounting for their resistance to certain pests. Creosote, an oily brown liquid One of two immediately cubital vein obtained from the distillation of wood tar, is widely used for dipping fenceposts, electrical and telephone poles, and wood in foundations for protection against termites and chinch bugs. It also acts as a fungicide and it is very toxic to plant. In 1976 the Environmental Protection Agency placed creosote on its "may be too hazardous to man or the environment" list, meaning that its use could be deregistered or severely restricted. Pentachlorophenol, on the same list, has been widely used against termites on poles and foundation timbers. Repellents for insects that bother livestock exist but require repeated applications, some at least twice a day, because they are rapidly absorbed through the skin of the animal. People devoted to outdoor sports, outdoor workers, and gardeners are familiar with the unusual protection provided by commercial insect repellents. Diethylm-toluamide ("deet") is the active ingredient in off and Cutterrs. 2-Ethyl 1, 3-hexanediol (6-12) is active in other preparations. Both give good, long-lasting, safe protection against mosquitos, flies, and fleas. Benzil, benzyl benzoate and dibutyl phthalate protect against mites (chiggers).

Reproductive potential: The maximum possible rate of increase (multiplication) of an organism.

Residual activity: The insecticidal activity resulting from the presence of an insecticide and its metabolites on or in a treated site after an insecticide is applied.

Residual deposit: Amount of insecticide remaining on a surface after application

Residual pesticide: A pesticide chemical that can destroy pests or prevent them from causing disease, damage, or destruction for more than a few hours after it is applied.

Residue: 1. The amount (often a trace) of pesticide and its metabolites left in food crop after treatment. 2. It is that part of a pesticide which remains in or on a plant or animal after application. It is expressed as ppm and is based on fresh weight of the sample.

Apart from the residues which result from the use of pesticides on food crops, there are also unintentional residues (residues which occur in food as a result of circumstances not designed to protect the food against the attack of pests). Such residues may occur (i) in crops grown in previously treated soil and (ii) in animal products (milk and its products, meat and eggs) as a result of feeding of animals on fodder crops which have earlier received pesticidal treatment. (iii) Amount of pesticide remaining in or on plant tissues (or in soil) after a given time, especially at harvest time.

Resistance: 1. The ability of an organism to suppress or retard the injurious effects of a pesticide. 2. Of insects to insecticides, the ability in a population of insects to tolerate a dosage of a pesticide which kills majority of individuals in a normal population of the same insect species. Of plants to insect attack, the ability of certain varieties to produce high yields of good quality as compared with those of ordinary varieties, at the same level of insect infestation. 3. The relative amount of inherited qualities which allow an organism to influence or reduce the damage done to it by its enemies. 4. Ability to avoid or repel attack by pest (or parasites): capacity to withstand toxic effects of pesticide. 5. The natural or induced capacity to avoid or repel attack by pests (or parasites). Also the ability to withstand the toxic effects of a pesticide or a group of pesticides, often by metabolic detoxification. Ability of an organism to tolerate the poisonous effect of a toxicant.

Resurgence: An increase of a (pest) population, after a period of decrease, to a level higher than its original one, especially when the decrease is caused by an intended control operation.

Retina: 1. The receptive apparatus of an eye. 2. The part of an eye upon which the image is formed.

Retina cells: The cells composing the retina.

Retinal pigment cells: Pigment cells in the retinal region on the eye.

Retinula: A group of two, three, or more light sensitive cells, which surround and secrete a longitudinal light receptor unit (optic rod or rhabdom).

Rhabdom: The light receptor unit or rod lying in the axis of a group of light-sensitive cells (retinula) below the crystalline cone of an eye; retinal rod; optic rod.

Rhabdomere: 1. One of 7 to 8 segments making up a retinal rod (rhabdom). 2. The receptive area of a retina cell that is one of the component parts of a rhabdom.

Rickettsia: An obligate intracellular parasite of arthropods, many types of which are pathogenic to human beings and other animals. They are thought to be intermediate between bacteria and viruses, because they have features in common with both.

Rostrum: 1. The beak or proboscis of Hemiptera. 2. The extended fore part of the head of an insect consisting of or bearing the mouth-parts. The term is applied to insects such as weevils in which biting mouth parts are at the tip of snout as well as to insects with piercing/sucking mouth parts.

Royal chamber: The space reserved for the accommodation of king and queen termites, or the space reserved for the queen of an ant colony.

Royal jelly: 1. Also called bee-milk. A special food given by worker bees to the larvae which are to be developed as queen bees. 2. A material supplied by workers to female larvae in royal cells which is necessary for the transformation of larvae into queens. Royal jelly is secreted primarily by the hypopharyngeal glands and consists of a rich mixture of nutrient substances, many of them possessing a complex chemical structure.

Runoff: 1. The sprayed liquid which does not remain on the plant. 2. The process of spray shedding from a plant surface during and immediately after application, when droplets coalesce to form a continuous film and surplus liquid drops from the surface.

Salivarium: The pocket between the base of the hypopharynx and the labium into which opens the salivary duct; in higher insects converted into a salivary pump or a spinning apparatus.

Salivary glands: Glands that open into the mouth and secrete a fluid digestive, irritant, or anticoagulatory properties.

Saltatorial legs: Legs adapted for jumping.

Scape: Basal segment of insect antenna; leafless pend uncle arising from a plant and bearing flowers.

Scarabaeiform larva: A grub-like larva, cylindrical, curved and with short thoracic legs but no prolegs.

Scavenger: An animal that feeds on dead plants or animals. on decaying organic matter, or on animal wastes.

Scent gland: A gland producing an odourous substance.

Sclerite: A plate of the body wall of an insect, bordered by sutures or membranes.

Scleroma: The sclerotic annulus of a body segment in distinction to the membranous conjunctiva.

Sclerotin: The substance or substances forming the hard parts of the insect's skeleton and deposited in the chitin.

Sclerotization: The hardening and darkening processes in the cuticle (involves the epicuticle and exocuticle).

Sclerotized: Hardened by deposition or formation of certain substance in the cuticle.

Scolytoid larva: A fleshy larva resembling the larva of a scolytid beetle.

Scutellum: The area of the alinotum posterior to the suture of the V-shaped notal ridge, or the corresponding area when the ridge is incomplete or absent.

Scutoscutellar suture: The external suture of the V-shaped notal ridge of the alinotum, the arms divergent posteriorly, dividing the notum into scutum and scutellum.

Scutum: In ticks, the sclerotized plate covering all or most of the dorsum in males, and the anterior portion in females, nymphs, and larvae of the Ixodidae.

Sebaceous gland: A gland producing a greasy lubrication substance.

Second axillary: The pivotal plate of the wing base resting on the pleural wing process, connected with the base of the radial vein.

Second median plate: A variable sclerotization at the base of the mediocubital field, folding convexly on the outer edge of the first median plate along the plica basalis; often absent, or represented by the united bases of the medial and cubital veins.

Second spiracle: The metathoracic spiracle, located near the anterior margin of the metapleuron, between the mesopleuron and the metapleuron, or in the posterior margin of the mesopleuron.

Secondary parasite: A parasite which establishes itself in or upon a host that is a primary parasite.

Secondary parasitoid: An insect parasite of a primary parasitoid.

Secondary segmentation: Any form of body segmentation that does not strictly conform with the embryonic metamerism; the usual segmentation of arthropods having a well developed exoskeleton, in which the membranous intersegmental rings are the posterior parts of the primary segments.

Sectorial cross vein: A cross vein connecting two branches of the radial sector Segment. 1. A ring or subdivision of the body or of an appendage between areas of flexibility associated with muscle attachment. 2. A body segment is any of the successive annular subdivisions of the arthropod trunks, whether corresponding to the embryonic metameres or produced secondarily. 3. A subdivision of the body or of an appendage, between joints or articulations.

Segmental appendages: The paired ventrolateral segmental outgrowths of the body wall serving primarily for locomotion.

Segmentation: 1. The body of insects and other arthropods is divided into a number of segments. 2. The embryological process by which the insect body becomes divided into a series of parts or segments.

Seminal vesicles: Pouch-like structures in which seminal fluid of the male is stored.

Semio-chemicals: Substances that modify behaviour of insects by inducing in them the response for sex, feeding, fear or alarm etc. These are further classified into pheromones, secreted by insects for stimulating the insects of the same species (i.e. intraspecific stimulators) and allelochemicals, the substances used for interspecific communications.

Sense organ: Any specialized innervated structure of the body wall receptive to external stimuli, most insect sense organs are innervated setae.

Sensillum: A simple sense organ, or one of the structural units of a compound sense organ.

Sensory nerves: Afferent nerves carrying sensory information to the central nervous system. Most of the nerve poisons (e.g.

organophos phate and organocarbamate insecticides) cut off this information system resulting in paralysis of the organism.

Sensory neuron: A neuron leading from a receptor cell to the central nervous system.

Septicemia: A morbid condition caused by invasion and multiplication of microorganisms in the blood.

Sericulture: The raising of silk cocoons, the basic material for silk production.

Serosa: The dorsal blastoderm or its extension as the outer layer of the amniotic covering of the embryo.

Serrate antennae: Antennae having a row of teeth, like that of a saw.

Setaceous antenna: Antenna that tapers towards the tip.

Sex attractant: A volatile chemical substance produced by one sex of an insect (mostly female) to attract the opposite sex.

Sex pheromone: A substance secreted by one sex of an insect for attracting the opposite sex of its own species to facilitate mating.

Sex ratio: The ratio of males tofemales in a population.

Shelf life: It is a period of time during which the pesticide may be stored without becoming unfit for use; storage life.

Sibling: One of two or more individuals having common parents.

Silk gland: Gland in the silkworm and other insects secreting the substance produced as silk.

Skeletal muscle: In insects, a muscle that stretches across the body wall and serves to move one segment on another.

Skeleton: The sclerotized (hard) parts which form the external protective covering of the body and the internal points of attachment for the muscles.

Social homeostasis: The maintenance of steady states at the level of the society either by control of the nest microclimate or by the regulation of the population density, behaviour, and physiology of the group members as a whole.

Soil fumigant: A pesticide that will evaporate quickly. When added to the soil, the gas formed kills pests in the soil. Usually a tarpaulin, plastic sheet, or layer of water is used to trap the gas in the soil until it does the job.

Soldier: 1. A member of a worker subcaste specialized for colony defense. 2. In termites, sterile males or females with large heads and mandibles; they function to protect the colony.

Solid cone: Jet with air-core reduced to give a cone of spray droplets.

Solid cone nozzle: A spray nozzle similar to the hollow-cone nozzle, except that it emits a solid cone of spray; especially suited for close spraying, as with boom spraying of field crops.

Solitary parasitoid: An insect parasite which normally develops at a rate of one individual per arthropod host.

Somatic cells: The body cells as distinguished from the germ cells.

Speciation: Separation of one population into two or more reproductively isolated independent evolutionary units.

Species: 1. A group of individuals or populations that are similar in structure and physiology and are capable of interbreeding and producing fertile offspring; the largest group of individual within which effective gene flow occurs or could occur. 2. Groups of actually (or potentially) inter-breeding natural populations which are reproductively isolated from other such groups.

Sperm cyst: One of the cellular capsules in the testis containing the spermatocytes.

Sperm tube: One of the secondary divisions of the testis.

Spermatheca: 1. The sperm storage receptacle of the female insect. 2. Receptacle in the female which receives and stores the sperm of the male.

Spermathical gland: A gland associated with the spermatheca in the reproductive system of female insects.

Spermathophase: A package enclosing sperm which is transferred to the female during copulation.

Spermatid: An immature spermatozoon in which the lagellum has not yet developed.

Spermatocyte: The sperm cell differentiated from a spermatogonium, before maturation.

Spermatogonium: The first stage in the differentiation of a sperm cell from a primitive male germ cell.

Spermatophore: 1. Gelatinous covered packets containing sperm. 2. A sperm case.

Spermatozoon (pl. spermatozoa): The mature male sexual cell or sperm cell, whose function is the fertilization of the egg.

Spermiogenesis: Process of transforming spermatids into functional sperm cells.

Spinasternum: One of the spina-bearing intersegmental sclerites of the thoracic venter, associated, or united, with the sternum preceding; a spinasternum may become a part of the definitive prosternum or mesosternum, but not of the metasternum.

Spinneret: A structure from which silk is spun.

Spinning glands: The glands which produce viscous secretion forming silk.

Spiracle: 1. It is one of the several small openings in the segments of insects connected to a system of internal tubes through which oxygen is inhaled and carbon dioxide is exhaled. 2. Small openings in the abdominal segment of insects connected to a system of internal tubes (trachea) through which oxygen and carbon dioxide are exchanged with the atmosphere. 3. An external opening of the tracheal system through which diffusion of gases takes place.

Spray low volume (LV): The technique requiring 50 to 100 litres of liquid formulation per hectare coverage accomplished with a mist blower. It requires lesser time and reduced labour cost as compared to high volume sprays.

Spray nozzle: The attachment on the end of the sprayer extension that breaks up spray liquids into fine drops so as to apply the material evenly on the plants.

Sprayer: A machine used to apply a spray by forcing the liquid through a nozzle under pressure. Field crop sprayers are classified in terms of their capacity for application, as follows:

	Volume	
	litres/ha	**gal.ac.**
Low	less than 220	less than 20
Medium	220-260	20-60
High	more than 660	more than 60

Sprayers essentially consist of a tank from which the liquid is pumped to the nozzles which are fitted at regular intervals along a boom. Nozzles may be of the fan type (used for low and medium volume spraying) or cone type (used for high volume spraying). Sprayers with tanks up to 450 I (c. 100 gal.) capacity may be tractor mounted. Those of larger capacity, up to 1800 I (400 gal.), are either trailed or mounted on 4-wheel drive vehicles. Common types of spray nozzle produce a wide range of droplet sizes. In recent years sprayers have been developed which use high-speed spinning nylon discs with fine-toothed margins to produce droplets of uniform size (Controlled Droplet Application-CDA) in the range 250-300 microns. These sprayers use less spray, give good crop cover and produce less SPRAY DRIFT. Other sprayers designed to apply small volumes of liquid (generally less than 5 litres/ha) (Ultra Low Volume-ULV) use a droplet size of 60-70 microns which rely on being carried in the wind to give thorough crop cover. ULV sprayers are prone to spray drift and are not used to apply herbicides or poisonous chemicals. more than 60 Electrostatic ULV sprayers are nowbeing developed. These produce positively charged droplets of controlled size (c.50 microns) which behave fairly predictably and are attracted to the mainly uncharged target crop. Spraying may also be carried out by aeroplane or helicopter.

Springing organ: In Collembola, the organs help in jumping.

Stadium (pl: stadia): The time interval between molts in a developing insect.

Sterile male technique: A method of pest control which involves the release of large number of sterile male insects in the populations.

Sterile-insect technique: A genetic method of pest suppression involving the mass release of compatible but sterile insects into a wild fertile pest population to overwhelm and suppress its reproductive capacity, often eventually to the point of extinction.

Sternacosta: The transverse internal ridge of the sternal suture through the bases of the sternal apophyses.

Sternacostal suture: The external suture of the sternacosta, separating the basisternum from the sternellum.

Sternal apophyses: The lateral apodemal arms of the eusternum; in higher insects united on a median base, the whole structure forming the furca.

Sternellum: The area of the eusternum posterior to the bases of the sternal apophyses of the sternacostal suture.

Sternite: A ventral sclerite, in other words, a portion of the body wall bounded by sutures and located in a ventral position (opposed to TERGITE).

Sternopleural bristles: Bristles on the sternopleuron (Diptera).

Sternopleurite: The infracoxal sclerotization of a generalized thoracic pleuron, generally united with the primary sternum in the definitive eusternal plate.

Sternopleuron (pl:, sternopleura): A sclerite in the lateral wall of the thorax, just above the base of the middle leg (Diptera).

Sternum: The ventral part of a body segment.

Stomach: The principal digestive sac, usually a muscular enlargement of the gut having a lining of glandular cells which secrets enzyme.

Stomodaeal nervous system: The nervous system centering in the ganglia of the stomooaeum. (Stomatogastric, visceral, or sympathetic, nervous system).

Stomodeum: The forgut of an arthropod.

Storage excretion: It is the temporary or permanent accumulation of excess of waste products, not dealt by the usual excretory organs, in special cells called the urate cells in the fat body.

Striated border: The inner cytoplasmic layer of the ventricular epithelial cells with fine lines perpendicular to the surface. (Stabchensaum, Harchensaum, plateau strie bordure en brosse.)

Striated muscle: Muscle that is composed of fibers with alternate light and dark bands.

Stridulate: To make a noise (chirp or creak) by rubbing two structures or surface together.

Stridulation: The production of sound by rubbing one part of the body surface against another.

Structure-activity relationship: A biological method to analyze the action of pharmacologically active chemicals such as pesticides by utilizing their physicochemical characteristics (e.g. hydrophobicity, steric hindrance, intra molecular electron distribution etc.) to determine the requirements for optimum biological activity.

Stylet: 1. A hollow tube like organ, especially that possessed, by aphids with which they probe the tissues of plants and through which the sugary contents of the phloem vessels is transmitted to the insect. 2. A hollow tube like organ, especially that possessed by aphid through which the sugary contents of the phloem vessels are sucked by the insect.

Subalare: The epimeral epipleurite giving insertion to the posterior pleural muscle of the wing (Epimeralgelcnkstuck, post-parapteron).

Subantennal suture: A suture on the face extending ventrally from the base of the antenna.

Subcosta: The usual second vein of the wing, associated basally with the anterior end of the first axillary sclerite.

Subcoxa: A secondary proximal subdivision of the coxopodite present in some arthropods, forming a support in the pleural wall of the body segment for the rest of the appendage (pleuron).

Subesophageal commissure: The two nerves that pass around the esophagus and connect the brain to the subesophageal ganglion.

Subesophageal ganglion: The knot like swelling at the anterior end of the ventral nerve cord, below the esophagus.

Subgenal areas: The usually narrow lateral marginal areas of the cranium set off by the subgenal sutures above the gnathal appendages, including the pliurostomata and hypostomata.

Subgenal sutures: The lateral submarginal grooves of the cranium just above the bases of the gnathal appendages, forming internally a subgenal ridge on each side, continuous anteriorly with the epistomal suture when the latter is present.

Subgenital plate: In the female the eighth abdominal sternum, or the seventh when the eighth is reduced or obliterated.

Subgenus (pl:, subgenera): A major subdivision of a genus, containing a group of related species. In scientific names, subgeneric names are capitalized and placed in parentheses following the genus name.

Submarginal vein: A vein immediately behind and paralleling the costal margin of the wing (Chalcidoidea).

Submentum: The proximal sclerite of the labium, by means of which it is attached to the head.

Subocular suture: A suture extending ventrally from the compound eye.

Subspecies: A subdivision of a species, usually a geographic race and are capable of interbreeding.

Suctorial: Adapted for or capable of sucking, having sucker for feeding or adhering.

Superparasitism: A condition resulting from the use of a single host individual by more individual parasitoids of the same species than it can successfully sustain to maturity because of nutritional limitations.

Superposition eye: An eye which permits the passage of light through the non-pigmented wall of one unit (ommatidium) to the neighbouring visual units.

Superposition image: Image produced by superposition or overlapping of points of light from a number of facets.

Supervised control: The control of pests as soon as it becomes necessary so that the pest does not increase its density above the economic threshold. Such a control necessitates keeping a regular record of the pest populations.

Susceptibility: Means the degree to which an organism is affected by a pesticide at a particular level of exposure.

Susceptible: Capable of being injured, diseased, or poisoned by a pesticide; not immune.

Susceptible species: A plant or animal that is affected by moderate amounts of a pesticide.

Suture: A line indicating the division of distrinct parts of the body wall.

Swarm: A concerted departure or association of a large group of insects from a nest or breeding habitat.

Sympatric: Inhabiting the same geographic region.

Symptom: Warning that something is wrong. Any indication of disease or poisoning in a plant or animal. This information is used to figure out what insect, fungus, other pest, or pesticide is causing the disease, damage, or destruction.

Synapse: 1. The region of contact between processes of two adjacent neurons, forming the place where a nervous impulse is transmitted from one neuron to another. 2. The central mechanism of intercommunication between the terminal fibres of two or more neurones.

Synergism: 1. Increased activity resulting from the effect of one chemical on another. 2. Increased pesticidal activity of a mixture of pesticides above that of the sum of the values of the individual components. 3. An activity of two or more agents which is greater than would be expected from summation of their single actions. 4. When the effect of two or more pesticides applied together is greater than the sum of the individual pesticides applied separately.

Synergist: A nontoxic chemical used to increase the effectiveness of an insecticide.

Synonyms: Two or more names for the same thing.

Systemic. 1. A compound that is capable of absorption into plant sap or animal blood. The systemic pesticides in this way render the sap or blood toxic to those pests that feed in or on the treated host. 2. A term used to describe certain pesticides that function by entering and becoming distributed within the plant as opposed to the pesticides that function by contact of the plants surface. 3. A pesticide absorbed through the plant surfaces (usually roots) and translocated through the plant vascular system.

Tactile organs: tactile receptors. The organs of touch.

Taenidium (pl:, taenidia): A spiral or circular thickening of the inner surface of the tracheae.

Tagoreceptor: An organ of touch having a single sense cell.

Tarsal formula: The number of tarsal segments on the front, middle, and hind tarsi, respectively.

Tarsus: The foot of an insect, the one- to five segmented appendage attached to the tibia, or lower leg segment.

Taxis (pl:, taxes): A directed response involving the movement of an animal toward or away from a stimulus.

Taxon: A group of real organisms recognized as a formal unit at any level of a hierarchic classification.

Taxonomy: The theory and principles of classifying organisms.

Tegmen (pl:, tegmina): The leathery or parchment like forewings of insects, such as cockroach and grasshoppers.

Tegula (pl: tegulae): A small scale like sclerite at the extreme base of the costa of the fore wing (particularly in the Lepidoptera).

Telopodite: The primary shaft of the limb distal to the coxopodite, the basal segment of which is the first trochanter (basipodite).

Telotarsus: The distal of the two principal subsegments of the tarsus in Arachinida and Chilopoda.

Telson: The terminal body regment in some primitive arthropods.

Tendon: A slender chitinous plate, band or strap to which muscles are attached for moving appendages.

Tentorial pits: The external depressions in the cranial wall at the roots of the tentorial arms; the anterior tentorial pits (at) located in the subgenal sutures or usually in the epistomal suture, the posterior tentorial pits in the lower ends of the post-occipital suture.

Tentorium (pl: tentoria): The endoskeleton of the head, consisting of two or three apodemes.

Teratogenesis: Structural abnormalities of prenatal origin, present at birth or manifested shortly thereafter.

Tergite. 1. A dorsal sclerite in other words a portion of the body wall bounded by sutures and located in a dorsal position (opposed to sternite). 2. A tergal sclerite; the dorsal surface of an abdominal segment.

Tergum (pI: terga): Dorsal part of a body segment; often referred to as the notum in thoracic segments.

Terminal: At the end; at the posterior end (of the abdomen); the last of a series.

Terminal filament: The cellular end thread of an ovariole. (Endfaden).

Termitarium: A termite nest. Also, an artificial nest used in the laboratory to house termites.

Thelytoky: Parthenogenetic reproduction in which only female progeny are produced.

Third Axillary: The flexor sclerite of the wing base; the sclerite on which the flexor muscle is inserted.

Third generation pesticides: A term proposed by Williams (1967) for juvenoids and other natural or synthetic chemicals potentially useful for pest suppression by virtue of their effective interference with hormonally directed natural processes in insects.

Thoracic tergites: Sclerites (hardened plates) on the dorsal part of the thorax. Also called NOTA.

Thorax: 1. The middle part of an animal's or insect's body, between the head and the abdomen. 2. The second or intermediate region of the insect body, bearing the true legs and wings. It consist of three segments, the pro, meso, and meta-thorax (named in order from the front). In insects, the legs and wings are borne on the three segments which comprise the thorax.

Threshold: The lowest value of any stimulus, signal or agency which will produce a specified effect.

Threshold limit value: The concentration of a chemical, the exposure to which for a healthy worker for eight hours a day, day after day for a working life time, produces no harmful effect.

Threshold population, economic: The pest population (weed, fungus, insect etc.) at which the cost of chemical control is recoverable by extra yield obtained as a result of the timely treatment. (Even low pest population causes decrease in yield, but the cost of chemical treatment at that level may be much more than the incremental yield obtained.

Tibia (pl: tibiae): The fourth segment of the leg, between the femur and the tarsus; often has strong spines.

Tibial spur: A large spine on the tibia, usually located at the distal end of the tibia.

Tolerance: Resistance of a pest to a pesticide, ability to survive normal dosage of a toxicant.

Tolerance level, median: The amount of a chemical in parts per million in food (oral tolerance) or in the environment (contact tolerance) which a sensitive species of an animal can tolerate for a short duration (4-7 days), without causing any noticeable harmful effect or injury. For example, TLM value for DDT against rainbow trout (Salmo gardener) is 0.01 ppm in water.

Tolerance level, pesticide: Pesticide tolerance has been defined by FAO and WHO as the maximum concentration of pesticide residue that is permitted in or on food at a special stage in the harvesting, storage or transport, marketing or preparation of food, up to the final point of consumption expressed in parts by weight of the food (ppm). These residue levels must be a result of good agricultural practices, toxicologically acceptable and be shown to present no hazards throughout the life of man or animal.

Tolerance limit: It is the maximum amount of a pesticide residue (ppm) that is permitted by regulation to remain in or on a commodity.

Tormogen: The epidermal cell associated with a seta which secretes the setal membrane or socket.

Toxicity: The inherent poisonous potency of a material.

Toxicology: The branch of science dealing with the study of poisons.

Trachea: A tube of the respiratory system that begins with the spiracle and end with the tracheoles.

Tracheal commissures: Transverse tracheal trunks continuous from one side of the body to the other.

Tracheal gills: Present in most aquatic insects, filiform or lamellate structures well supplied with tracheae and tracheoles.

Tracheal orifice: The primary opening at the point of formation of a trachea, whether exposed externally or concealed in a secondary a trial depression of the body wall.

Tracheal system: The part of the respiratory system composed of the tracheae and tracheoles.

Tracheoles: The terminal finer branches of the tracheae without spiral thickenings.

Trail pheromone: A substance laid down in the form of a trail by one animal and followed by another member of the same species.

Transverse costal vein: A cross vein in the costal cell (Hymenoptera).

Transverse cubital vein: A transverse vein connecting the marginal and cubital veins (Hymenoptera).

Transverse marginal vein: A cross vein in the marginal cell (Hymenoptera).

Transverse median vein: A cross vein between the median or discoidal vein and the anal vein (Hymenoptera).

Transverse suture: A suture across the mesonotum (Diptera).

Trap crop: A small area of a crop used to divert past from a larger area of the same or another crop.

Trichogen: An epidermal cell that generates a seta.

Trochanter: Each of the several bony processes on upper part of thighbone; second joint of an insect's leg.

Trochantin: The basal part of the trochanter when the latter is two segmented in the insect leg.

Trophallaxis: The exchange of alimentary liquid among colony members and guest organisms, either mutually or unilaterally. In stomodeal trophallaxis the material originates from the mouth; in proctodeal trophallaxis it originates from the anus.

Trophic: Pertaining to nourishment or feeding e.g. eutrophic means well fed or full nutrition.

Trophic level: Placement or grouping of all organisms which secure their food at a common step away from the first level e.g. (i) plants, (ii) herbivores, (iii) carnivores.

Trophocytes: Cells that elaborate nutritive materials; the cells of the fat body having a trophic in distinction from the urate cell.

Tympanal organs: Auditory organs on the abdomen or legs of certain insects.

Ultra low volume: Pesticide sprays at 0.6 to 4.7 liters per hactare or applied in an undeluted low volume formulation.

Unisexual: Of the sex only, e.g., aphids where asexual (parthenogenic) females are known.

Univoltine: Having only one complete generation annually.

Unsclerotized: Not hardened by the deposition of sclerotin.

Vagina: The terminus of the female reproductive tract that opens to the out side.

Valvifers: The basal plates of the ovipositor, probably from the coxopodites of the gonopods, carrying the valvulae, including first valvifers (1 *Vlf*) of the eight abdominal segment, and second valvifers (2*VIf*) of the ninth segment.

Valvulae: The three pairs of processes forming the blades and ensheathing lobes of the ovipositor. The first and second valvulae (1 VI, 2 VI) are gonapophyses of the gonopods; the third valvulae (3 VI) are distal outgrowths of the coxopodites of the ninth abdominal segment.

Venation: The arrangement of veins in the wings of insects.

Ventilation tracheae: Tracheae with collapsible walls, responding to varying surrounding pressure.

Ventral: Pertaining to the underside of the body.

Ventral diaphragm: A membranous and muscular sheet present in some insects stretched between the lateral edges of the abdominal sterna, sometimes extending into the thorax, separating the ventral sinus from the perivisceral sinus.

Ventral nerve cord: The chain of connected ventral ganglia, morphologically beginning with the tritcerebral ganglia of the brain; in entomology the term usually applied to the thoracic and abdominal ganglia only.

Ventral sinus: The space of the definitive body cavity below the ventral diaphragm, containing the nerve cord.

Ventral trachea: The ventral segmental trachea originating at a spiracle.

Ventral tube: In collembolan, a tube or tubercle proceeding from the ventral side of the first abdominal segment.

Ventriculus: Midgut; mesenteron or stomach of an insect.

Vermiculate: With irregular sinuous markings (worm-eaten).

Vermicule: Worm-like; very small worm.

Vermiform larva: A legless (apodous), headless (acephalic), worm-like larva typical of some diptera.

Vermin: Pests; usually rats, mice, or insects.

Vertbrates: Animals with a spinal column or backbone, such as fishes, birds, mammals, and so on.

Vertex: The top of the head, between the eyes and anterior to the occipital suture.

Vesicula seminalis: A dilation of the vas deferens in which the spermatozoa may be retained.

Vestigial: Having the nature of a degenerate or atrophied organ, more fully functional in an earlier stage of development of the individual or species.

Virulence: The disease-producing power of a microorganism, i.e. the relative capacity of a microorganism to invade and injure the tissues of its host.

Virus: A sub-microscopic pathogen that requires living cells for growth and is capable of causing disease in plants or animals. Plant viruses are often spread by insects.

Viscera: The internal organs in the cavity of the body particularly those in abdomen.

Visceral muscle: A muscle which invests an internal organ.

Visceral trachea: The median segmental trachea originating at a spiracle, branching to the alimentary canal, the fat tissue, and the reproductive organs.

Visceral tracheal trunk: A longitudinal tracheal trunk closely associated with the walls of the alimentary canal.

Vitellarium: The part of an ovariole egg tube in which the oocytes grow by the accumulation of yolk and attain their mature size. (Zone of growth.)

Vitelline membrane: The wall of the egg cell; undivided in meroblastic cleavage, surrounding the blastoderm.

Vitellophags: Endoderm cells proliferated into the yolk and accomplishing a partial digestion of the latter.

Viviparous reproduction: Type of reproduction in insects, where embryo develops within the body of the insect and young ones produced similar to adults; found in some flies.

Vulva: The external opening of the genital chamber or vagina serving in most cases for both copulation and the discharge of the eggs; sometimes on the eighth abdominal segment, sometimes on the ninth.

Waggle dance: The dance whereby workers of various species of honeybees (genus Apis) communicate the location of food, finds and new nest sites. The dance is basically a run thriugh a figure eight pattern, with the middle, transverse line of eight containing information about the direction and distance of of the target.

Waiting period: It is a time limitation in respect of harvest of the treated crop or slaughter of the treated animal; safety period.

Wettable powder: Pesticide formulation of active ingredient mixed with inert dust and a wetting agent, which mixes readily with water and forms a short-term suspension.

Wetting agent: A substance which lowers the surface tension of water so that it spreads out rather than remains in droplet form when sprayed. Wetting agents are added to detergents to increase their efficiency and to sprays of insecticides, fungicides, herbicides or mineral sprays to improve the cover achieved when applied to leaf surfaces.

Wildlife: All living things that are not human or domesticated, nor pests, as used here, including birds, mammals, and acquatic life.

Wing base: The proximal part of the wing between the bases of the veins and the body, containing the humeral and axillary sclerites.

Wing pads: The undeveloped wings of nymphs and naiads. They appear as two flat structures on each side.

Xenobiosis: The relation in which colonies of one species live in the nests of another species and move freely among the hosts, obtaining food from them by regurgitation or other means but still keeping their brood separate.

Xenophagous: Feeding on non-specific host plant.

Xylophagous: Of insects which feed on wood.

Zero tolerance: By law, no detectable amount of the pesticide may remain on the raw agriculture commodity when it is offered for shipment. Zero tolerances are no longer allowed.

Zooplankton: Microscopic animal living suspended in water.

❑❑❑

3 Horticulture

Abort: To stop or fail to reach full development (maturity) or to develop imperfectly; as some hybrids are prone to embryo abortion or as some attempts to culture a plant may fail (abortive attempt).

Abscisic acid (ABA): A plant growth substance associated with leaf senescence, fruit drop, seed and bud dormancy, some aspects of apical dominance and inhibition of flowering in long day plants under short-day conditions.

Absciss: The dropping of leaves, flowers, fruits or other plant parts, usually following the formation of an abscission zone or separation layer which has a role in the separation (abscission) of that plant organ from the plant body. The balance between auxins and abscisic acid may control abscission. Gibberellins have also been implicated.

Abscission layer: It forms at the base of the leaf stalk on deciduous trees, seals of the sap flow, producing a change in the colour of leaf and eventually falls.

Absiccsion zone: The region at the base of some plant organs consisting of a layer of weak cells and a protective layer of croak cells which protect the wound when the organ falls off.

Absorbance: A measure of the extent to which light is attenuated during passage through a coloured liquid or solid.

Absorbent: A substance used to absorb another.

Acaulescent Acauline: Stem less or having a very short stem.

Accessory bud: An additional bud; a bud in addition to the auxiliary bud.

Accessory fruit: A fruit formed not only from the ovary but also from other or accessory parts.

Acclimatization: Process under which plants are accustomed to a new environment with the aid of human management.

Accrecent: Increasing in size with age; used to plants that continue to grow after flowering.

Acer pseudoplatanus: The sycamore tree; a source of heterotrophic and autotrophic plant cell suspension cultures which have been widely used for research purposes.

Achene: Small single-seeded dry fruit which does not split ex.sycamore.

Achlamydeous: Flowers that lack petals and sepals, ex. Willow (Salix)

Achlorophyllous: Lacking chlorophyll

Achroous: Without colour, unpigmented. Achromatic.

Acicular: Needle-shaped or Aciform.

Acidotrophic: Feeding on acidic food or acidic substrates.

Acme: A prior of maximum vigour; the highest point.

Acne: A pimple-like eruption of the sebaceous glands of the skin, with accumulation of yellow secretion and black overgrowth of the horny layer of the skin.

Acotyledon: Plant without obvious cotyledons or seed leaves, ex.moss.

Active Ingredient: That part of a growth regulator formulation from which the biological activity is obtained.

Active site: The region of the enzyme with which the substrate binds during catalytic conversion to a product. The basic theory of enzyme activity assumes formation of an enzyme-substrate complex through binding at the active site. Electrostatic attraction, repulsion and other intermolecular physical forces contribute to a lowering of the activation energy, thus facilitating intermolecular conversions.

Adnate: Attached to another organ, especially down the entire length.

Adnation: In flowers, the congenital growing together of unlike parts, such as two or more whorls of floral parts.

Adsorption: The retention of one substance (the adosrbate) at the surface of another (the adsorbent).

Aduncate: Bent; hooked; crooked.

Adeventitious embryony: The production of an embryonic sporophyte by mitotic divisions from tissues of another sporophyte without an intervining gametophyte generation; seed production without a sexual process.

Adventitious root: Any root which grows from a tissue other than the pericyle or endodermis of an older root. These roots appear in an unusual place or position, e.g, on the stem in tulips and onion. Some plant may even produce adventitious roots from leaves.

Adventive: Formed or growing in an adventitious manner.

Aecidiospore: A type of spore produced by a rust fungus prior to the production of uredospores.

Aestivation: The arrangement of the parts of a flower of leaves within the bud.

Afforest: To plant with trees, to create a forest.

After-grass: Grass that grows among the stubble after has been harvested.

Agamic: Form of vegetative propagation whereby plant may be multiplied by division or by cuttings, as opposed to seminal propagation from seed.

Agamospermy: Any type of apomixes other than vegetative propagation.

Agamotropic: Used to flowers that do not close again one they have opened.

Age: 1. The period in the life cycle of an organism; the process of growing older, Mature. 2. The state of being old or senescent. 3. Culture age is a function of the number of subcultures and the time after subculture.

Agenesis: The absence of development.

Agglutinate: 1. To cause to unite, adhere; as if glued. 2. To gather into a clump or mass; as protoplasts and bacteria, in the presence of specific antibodies, tend to stick to one another. The process is agglutination.

Agglutinin: An antibody capable of clumping bacteria or other cells.

Aggregate fruit: A fruit consisting of a number of similar small fruitless all derived from the ovaries of a single flower, as in blackberries.

Agitate: To move a solution with an irregular rapid motion; as the contents of fermentors or flasks on shakers in order to maximize tissue or cell exposure to nutrients, facilitate gaseous exchange (aeration) and to disperse cells.

Aglet: Dangling or pendulous catkin, of birch, hazel, etc.

Agrobacterium tumefaciens: The bacterium causing crown gall disease of plants, inducing tumors to form. Tumors may be cultured in vitro and are useful in the study of this disease. The Ti plasmid of A. tumefaciens is known to cause the disease and a small portion of it is used as a vector in the genetic modification of higher plant cells. Novel DNA sequences are spliced into the plasmid DNA segment, the DNA is circularized and introduced into the cultured plant cells.

Air-layering: A method of propagating certain types of plant such a holly, lilac etc. A portion of stem is wrapped in damp moss or similar material and sealed in a waterproof membrane such as plastic. The moss must be kept moist, and rooting may take up to two years or more with some subjects.

Alar: Wing-shaped, wing like.

Alate: Winged; having wings or wing-like appendages, as the seeds of sycamore.

Albido: The white tissue beneath the peel of citrus fruits.

Albino: An organism lacking normal pigmentation, due to genetic factors; as in the white portion of variegated leaves. The condition is albinism. A conspicuous plastome (Plastid) mutant involving loss of chlorophyll; sometimes a product of plant tissue culture.

Alcove: Recess in a garden hedge or wall.

Aldicarb (Temik): Systemic insecticide.

Aleurone layer: The outermost layer of the endosperm.

Allodiploid: A hybrid diploid in which one or more chromosome pairs are derived from different species.

Allogamy: The production of zygotes by cross fertilization; allogamous.

Alpine: Originally a mountain plant but now loosely applied to plants suitable for rock gardens and similar conditions.

Alpine house: A greenhouse usually unheated specially ventilated and equipped mainly for the cultivation of alpine subjects.

Alternate host: Either of two different host plants of a pest or pathogen that requires both to complete its life cycle.

Alternate: Branches or leaves attached to the main stem singly at different heights and on different sides of the stem.

Alternation of generation: The occurrence in a life cycle of two or more different forms in successive generations, the offspring being unlike the parents and usually reproducing by a different method. It is a feature of ferns where the large spore bearing generation alternates with the smaller sexually reproducing generation.

Amphibious: Capable of growing in either water or soil. Watercress is such a plant.

Amphicryptophyte: A marsh plant with amphibious vegetative parts.

Amphidiploids or amphiploid: An interspecific hybrid with one of each kind of parental chromosomes per cell. Usually sterile.

Amphigean: Used of a plant having underground as well as aerial flowers.

Amphiphloic: Having phloem on both sides of the xylem.

Amphistomatic: Leaves having stomata on both surfaces.

Amphitoky: Parthenogenisis in which both male and female offspring are produced; ampherotoky; amphoterotoky; deuterotoky; gametotoky.

Anatropous: Having an ovule turned over to lie alongside its funicle (stalk).

Anbury: A disease of some root crops such as turnips.

Anemophilous: Wind-pollinated

Aneuploid: Individual cell or organism in which genetic change has occurred resulting from addition or subtraction of individual chromosomes and occasionally compounded by chromosomal rearrangements. The condition is aneuploidy or heteroploidy. Monosomics, trisomics and tetrasomics are aneuploid.

Angiocarpous: Having the fruit enclosed is a special casing, ex. Acorn.

Annelid: A member of the Annelida, the major group of segmented worms which includes the earthworm.

Annual: Plant that grows, fruit flowers and produces seeds in the course of a single year.

Anomalous: An exception to the rule, type or form; as irregular or abnormal growth or performance in cultured plants as compared to controls.

Anther culture: Refers to culture of single pollen grains or of the anther, containing the male gametophytes or micropores, with the objective of producing monoploid plants.

Anther Sac: The part of the anther containing the pollen.

Anther: That part of the stamen carried at the top of the filament and bearing the pollen.

Antheridium: The male cell in cryptogams that produces sperm; corresponds to the another in flowering plants.

Antiauxin: A chemical that interferes with the auxin reponse. They may or may not involve prevention of auxin transport or movement in plants. Some said to promote morphogenesis in vitro; as 2,3,5-triiodobenzoate (TIBA mw 499, 81), or 2, 4, 5-trichlorophenoxyacetate (2,4,5-T mw 255,49), which stimulate the growth of some cultures.

Antioxidants: A substance (such as ascorbic acid, citric acid or others) which is sometimes added to inhibit or prevent oxidative browning of the culture medium, due to bleeding or phenolic exudates. The latter may lead to tissue necrosis and death. Ascorbic acid (100 mg/l) and citric acid (150 mg/l) are most commonly used in sterilizing solutions.

Antipetalous: Opposite a petal.

Antisepalous: Opposite the sepals.

Antiseptic: The process or agent capable of inhibiting the growth of micro organisms.

Antiserum: A blood serum of an animal containing antibodies (to some specific antigen) and providing immunity from a specific disease.

Antrorse: Pointing forward or upward.

Aperturate: Having one or more openings, usually applied to pollen.

Apetalous: Lacking petals.

Apex: The most extreme point of growth of a plant; as the apical shoot and root tips are located at the apexes (apices), and contain the apical meristem.

Aphyllous: Lacking leaves.

Apical cell: A prominently differentiated meristem cell, derivatives of which organize the plant body or organ.

Apical dominance: The terminal bud influence, a widespread phenomenon in the plant kingdom, which is related to the auxin content of apical buds, and is exerted on lateral buds, resulting in their growth suppression. Abscisic acid may also inhibit lateral bud growth. Cytokinins tend to promote lateral bud development, thereby overcoming apical dominance.

Apical growth: Growth by means of an apical cell dividing to form the thallus beneath it.

Apical meristem: A region of actively dividing cells that occurs at the tip of roots and shoots.

Apposition: In plant cells the walls become thickened by the deposition of successive layers of cellulose.

Aroid: A plant belonging to the Arum Family and including anthuriums, dieffenbachias, monsteras and philodendrons.

Articulate: A natural joint or node where separation is easiest.

Artificial long days: Interruption of dark period or extension of natural day length to prevent flower bud initiation of short-day plants, such as chrysanthemums and poinsettias.

Artificial seed: Encapsulated or coated somatic embroyos (embryoids) that are planted and treated like seed.

Artificial selection: Plant selection by man for agronomic qualities.

Asexual: A non-sexual way to increase plants, such as by cuttings, layering and division rather than seed.

Asexual propagation: Reproduction by vegetative means, such as cutting, division.

Asexual reproduction: Development of new organisms without the fusion of gametes. This may occur in plants by either spore formation or vegetative reproduction.

Asparagine (Asn, $C_4H_8N_2O_3$ mw 132:12): An amino acid occasionally included in plants tissue culture media, as a source of reduced nitrogen.

Assay: 1. To test or evaluate. 2. The substance to be analyzed or the process of examining or testing it (Chemically or by other means).

Asymbiotic: A plant which is hostile to certain other plants, such as horse-radish (Cochlearia armorcia).

Atrinal (Dikegulac): Chemical pinching agent used to promote lateral shoot development on azaleas. Now sold as Atrimmec.

Attenuate: Gradually long-tapering, as of a leaf.

Atypical: Refers to an appearance of function not conforming in type.

Auricled. Having ear-like outgrowths, e.g, the lobes on some leaves.

Auriculate. Having an ear-shaped lobe or lobes, as in some leaves and petals.

Autecology: The study of an organism in relation to its environment.

Autopolyploid: A cell of individual with in excess of two of the monoploid chromosome sets (all homologous) that characterize the species. Those with an odd number of chromosome sets are usually sterile. Those with and even number of chromosome sets may have reduced fertility. The condition is autoploidy.

Auxin-cytokinin ratio: The relative proportion of auxin to cytokinin present in plant tissue culture medium. Varying the relative amounts of these two hormone groups in tissue culture formulas affects the proportional growth of shoots and roots in vitro. As the ratio is increased (increased auxin or decreased cytokinin contact), roots are more likely to be produced, and as it is decreased root growth declines and shoot initiation and growth are promoted.

Auxoautotrophic: Used of an autotrophic organism requiring certain specific additional nutrients for growth; auxoautotroph.

Auxoheterotrophic: Used of an organism that is unable to synthesize all substances necessary for growth and development.

Axile: Of that type of placentation in ovaries where two or more carpels unite at the centre, dividing the ovary into as many loculi as there are carpels.

Axilliary bud proliferation: Propagation in culture by protocol and media which promotes axillary (lateral shoot) growth. This is a technique for mass production (micro propagation) of plantlets in culture, achieved primarily through hormonal inhibition of apical dorminance and stimulation of lateral branching.

Axilliary bud: A bud located in the axil of a leaf.

Axilliary: Borne in an axil.

Axis: The whole main stem of a plant or the receptacle of a flower.

Azotobacter: A genus of nitrogen-fixing bacteria.

B-Nine (SADH) (N-dimethylamino) succinamic acid): Growth retardant use to control height or several floricultural crops and to promote flower initiation in other, such as azaleas.

Back bulb: A pseudobulb produced at the back of an orchid's flowering stem; one which has borne a flower in a previous season.

Back mutation: A mutation of a gene so that it assumes a function lost by an earlier mutation; reverse mutation.

Back-cross: Progeny derived from a cross between a hybrid and one of its parents.

Bactericide: A substance or agent that kills bacteria (usually rapidly); is bacterial.

Bacteriostat: A chemical or other agent that does not kill but prevents bacterial growth and multiplication; is bacteriostatic.

Bagging hook: A type of sickle or billhook.

Ball, E: (1946): The first to obtain plant from cultured shoot spices.

Ball-barrow: A type of wheelbarrow in which the wheel is replaced by an inflated plastic ball. Said to be easier to handle than a conventional barrow, especially across soft or rough ground.

Ballast: A step-up transformer that increases the voltage and decreases the amperage to lightning systems such as fluorescent, high pressure sodium and metal halide lamps. Ballasts generate a substantial amount of heat so may be isolated from the culture area in commercial operations.

Ballerina: Trade name for a type of fruit tree developed by East Mailing Research Station, in Kent; usually apple, but other are being developed, which grows as a single spike with fruit clustered down the entire length of the stem.

Bark graft: A graft made by slipping the scion beneath is slit in the bark of the stock.

Bark-bound: A condition in which the bark becomes too hard to expand in the normal way, thus strangling the tree. Lack of moisture in the soil is the most common cause. Occasionally the stress is relieved by the dark being split vertically, either naturally or by human action.

Bark: The dead corky cells on the exterior of stems and branches to protect trees etc. against predators.

Bastard trenching: Double digging.

Batch culture: A cell suspension grown in liquid medium of a set volume. Inocula or successive subcultures are of similar size and cultures contain about the same cell mass at the end of each passage. Cultures commonly exhibit five distint phases per passage; a lag phase follow inoculation, then an exponential growth phase, a linear growth phase, next a deceleration phase and finally a stationary phase.

Bedding plant: Plant used for temporary bedding displays, often raised under glass before being 'bedded out' in the open ground.

Bell glass: An early form of crop protection for tender subject such as lettuce. Extensively use in France in earlier times and very popular with English Victorian gardeners; now largely superseded by various types of cloche.

Benlate (Benomyl): Systemic fungicide

Benthos: Plants living on or attached to the sea bed; opposite to plankton which drifts on the surface.

Benzalkonium chloride or zephiran chloride: An alkyldimethylbenzylammonium chloride mixture sometimes employed as a disinfecting agent for plant material.

Benzylaminopurine: A synthetic cytokinin, also known as 6-benzyladenine.

Bergmann, L: (1960): Was first to obtain callus by plating cells from suspension cultures onto solid medium.

Berry: A pulpy fruit containing one or more seeds but no true stone, e. g, tomato, grape, gooseberry and raspberry.

Biauriculate: Having two auricules (ears), as of some leaves.

Bicarpellate: Derived from two carpels.

Bicephalous: Double-headed or having head in pairs.

Bicoloured: A plant having two colours on the same petal or in the same flower.

Biennial: A plant that lives through two growing seasons. It bears fruit and dies the second year.

Bifacial: Of leaves whose epidermis are alike.

Biferous: Bearing two crops of fruits in one season.

Bifid: Forked or split with two points.

Biflagellate: Bearing two elongated cytoplasmic extensions, the flagella.

Bifoliar: Having two leaves.

Bifoliate: Leaves consisting of two leaflets.

Bifurcate: Forked or Y-Shaped; divided into two branches.

Bigeneric hybrid: A plant produced by crossing two plants from different genera. This is indicated by a cross positioned in front of the plant's name. for instance, the ivy tree (x Fatshedera Liz) is a cross between a form of the false castor oil plant (Fastesia japonica 'Moseri' and also known as the Japanese fatsia), and the Irish ivy (Hedera helix 'Hibernica').

Bigeneric: Derived from two different generation hybrid.

Bilabiate: Having two lips, an upper and lower part, as in antirrhinums.

Bioassay: A biological assay or assessment procedure, performed on living cells or on a living organism; sometimes used to detect minute amounts or substances which influence or are essential to growth.

Biological control: The control of a pest by the introduction, preservation or facilitation of natural predators, parasites or other enemies, by sterilization techniques, by the use of inhibitory hormones or by other biological means; biocontrol.

Biological screen: A test or series of test using biological material to determine the effects of newly-synthesized chemical on growth.

Biont: An individual organism; biom.

Bionta: The ending of the name of subkingdom in botanical nomenclature.

Biotechnology: The industrial use of biological processes; as yeast fermentation for alcohol production or plant cell culture for extraction of secondary products. The scope of biotechnology has been dramatically increases through genetic engineering.

Biovulate: Having two ovules.

Bipartite: Divided into or consisting of two parts.

Biseriate: In two rows or two series, as a perianth consisting of a clayx and a corolla.

Biserrate: Serrate leaf with small teeth on larger teeth, as on silver birch.

Bisexual: Having both sexes in the same flower, hermaphrodite.

Biternate: Twice ternate; applied to leaves where the primary divisions are again divided into three parts.

Bleeding: The loss of sap from plant tissues that have been damaged or cut. Rubber trees are bled to extract the latex.

Blight: 1. General but incorrect term often used in reference to aphid infestation. 2. Type of fungus which attacks potatoes, tomatoes, etc.

Blind: A plant whose growing point has not developed properly.

Blind shoots: Shoots that remains vegetative under conditions that normally stimulate formation of inflorescences.

Blindness (flower): The failure of a bulb to produce any floral parts.

Bloom: 1. A flower or flower head. 2. The very fine powdery or waxy deposit covering the surface of certain leaves, stems and fruits such as grapes and plums; a protective coating easily rubbed off if the fruit is badly handled.

Blossom: 1. To open into flower. 2. A flower, or the mass of flowers on fruit trees, etc. in spring.

Bog garden: A type of garden where the soil is kept permanently very damp but not water logged.

Bole: The trunk of a tree between soil level and the first main branch.

Boll: Spherical seed-vessel of some plants such as cotton.

Bolting: Premature flowering. Under normal conditions all vegetable will eventually flower and set seed but most are harvested and used in their immature state. Some, however, e.g, lettuce will often bolt in hot weather, producing flower spikes but no hearts.

Bonsai: A method of dwarfing certain trees and shrubs, developed and favoured by the Japanese. It involves special shoot and root pruning coupled with complicated training systems.

Boscage: Wooded scenery; a dense area of trees or shrubs.

Bostryx: A cymose inflorescence in which the lateral branches developed on one side of the stem only, as found in some of the Hypericaceae family.

Botanize: To seek or collect plant for study.

Bottle garden: A specialized form of indoor plant cultivation. Many plants thrive in totally sealed glass containers, the moisture transpired condensing on the inner surface of the glass and being returned to the soil. Large carboys are excellent for this purpose and carefully planted; make ideal bottle gardens, or special garden centers. Bottle gardening Growing plants in environments created by carboys and other large glass jars. Sometimes the container is stoppered and the air inside recycled by plants; while others are left open.

Bottle graft: A plant graft in which the scion is a detached branched and is protected from wilting by keeping the base of the branch in a bottle of water until union with the stock.

Bottle: A bundle of hay or straw, sometimes reeds.

Bottom Heat: Gentle heat applied to the base of a cutting, seedling or plant to assist rooting or development. In the past animal manure was used to provide bottom heat but electric soil-heating cables are the modern equivalent.

Bramble: Coarse prickly shrub, usually invasive, e.g. blackberry.

Branch: Limb of a tree, shrub, etc., more substantial than a twig but less so than a bough.

Branchlet: Small branch, in size between a branch and a twig.

Brash: Hedge-clippings and similar refuse.

Brassica: A member of the cabbage family including cauliflower, turnip, swede, etc.

Bread fruit: Artocarpus altiliis. An Indo-Malaysian tree.

Bridge grafting: A method of grafting to overcome the total girdling of a tree either by disease, animals, accident or vandalism.

Bridge: A filter paper or other substrate used as a wick and support structure for a plant tissue culture when liquid media are used.

Broadcasting: The method of random sowing of seed by casting it freely on the prepared ground without recourse to pre-drilling or lining.

Brushwood: Small trees and shrubs, under wood.

Brutting: The fracturing- but not severing- of the young new shoots of fruit trees to restrict late growth.

Bud opening: Development of mature flowers in the wholesale or retail outlets; flowers cut prior to opening.

Bud scales: Leaf like structures that surround some flower buds, such as on azaleas.

Budding: A grafting method often used on roses and fruit trees. A well developed bud is taken from one plant and inserted under the bark of another. In this way a flower produced by a delicate plant can be grown on to a more robust rootstock.

Bulb forcer: A greenhouse operator who produces pot plants and cut flowers from bulbous species using artificial growing conditions.

Bulb grower (producer): An individual or firm who propagates bulbs for wholesale purposes.

Bulb maturity: Measure of capacity of a healthy daughter stem axis to sprout without delay and to respond to flower-inducing treatments.

Bulb production phase: All aspects of bulb production that lead to the sale of forcing sized bulbs. The production phase may take 1 to 3 years depending on the species.

Bulb programming phase: All temperature treatments given to the bulbs from the time they are harvested until they are placed under green house conditions.

Bulb: A food storage organ consisting usually (but not always) of an underground type of bud with a short thick stem and tightly packed, fleshy scales. Corms,tubers and rhizomes are often erroneously described as bulbs.

Bulbi: An un elongated axis, often with thickened phyllids, that serves in vegetative propagation in moss.

Bulbifery: Asexual reproduction by the formation of bulbils in the axils of leaves, or in the floral branches.

Bulbil: An immature and miniature bulb that usually grows at the base of another bulb. However, some plants, such as the mother fern (Asplenium bulbiferum), develop plantlets on their leaves which are also known as bulbs. These can be detached carefully using tweezers and encouraged to form roots.

Burgeon: To flourish, bud, sprout or grow.

Burgundy mixture: A general purpose fungicide consisting of a combination of ordinary washing soda and copper sulphate, a favourite with Victorian gardener but now largely superseded by proprietary products.

Burr: 1: Hooked or barbed fruit or flower head which attaches itself to a passing animal as a means of dispersal. 2. A woody projection or outgrowth found on the stems of some types of tree.

Bypassing shoot: Vegetative shoot that develops immediately below the flower bed, often occurring on azaleas.

Cabinet (growth cabinet): A cupboard suitable for incubating a small number of culture vessels or tubes under controlled environments. The degree or control over light, temperature or relative humidity is a function of quality of the cabinet but is generally less than that implied by the use of the term incubator or laboratory incubator Calcium alginate. A hydrophyllic colloid obtained from seaweed. A derivative of alginic acid $(C_6H_8O_6)n$. It reversibly immobilizes plant protoplasts in which condition they can tolerate a reduced osmotic potential.

Cactus: A member of the family Cactaceae, fleshy succulents whose stems store water and do the work of leaves, often carrying spines. There are about 200 genera and over 2000 species in cultivation. A characteristic of cacti is that they bear areoles.

Calcareous: Applied to soil containing a high measure of lime.

Calcicole: A lime-loving plant. Calcicolous plants grow best when lime is present in the soil. Limy soils are alkaline with a pH above 7.

Callosity: A callus or thickening; a protuberance.

Compatibility: Physiological and genetic possibility of pollination and of fertilization.

Campylotropous: Ovules bent over at 90 degree to the stalk which appears to be attached to the side of the ovule.

Cane brake: American term for a thicket or overgrown area of land.

Cane: Hollow or solid straight stem of certain plants such as some reeds and grasses, raspberries, etc.

Canescent: Covered in short, fine hairs, usually grey or white; grey hoary in appearance.

Canopy: The top layer of a forest consisting of the crowns of trees.

Carbon source: A source of the non-metallic element carbon (C) as organic substances like sugars taken up and metabolized by plant tissue cultures.

Caudicle: A stigmatic prolongation in the flowers or Orchids, uniting the two pollen masses.

Caulogenesis: Shoot formation; as de novo shoot development from callus.

Cell count: The number of cells per unit suspension volume or callus weight. Tissue is treated with chromic acid (5-8%) or pectinase (0.25%) for up to 15 minutes followed by mechanical dispersion, then cell numbers are estimated with a hemocytometer (haemocytometer).

Cell culture: The culture of single or groups of cells on solid or dispersed in liquid nutrient media (cell suspension cultures) usually following a set of defined growing conditions (protocols).

Cell hybridization: Formation of synkaryons, viable cell hybrids produced through cell fusion. They are identifiable by their increased chromosome number compared to the parent cells and the possession of characters found in one or another of the parental cells.

Cell line: Developmental history or decent, through cell division from a single original cell; as callus may constitute a group of cell; as callus may constitute a group of cell, all descendants from a single cell plating. Numerous cell lines may be present in culture. Any deviation in culture technique may favour one cell line over another.

Cell suspension: Cells and small aggregates of cells suspended in a liquid medium; as in cell suspension cultures. Explants, or callus derived from them are transferred to liquid medium and the cultures are then agitated on a mechanical shaker. The ensuing single cells and small cell clusters are used for a number of purposes in plants tissue cultures; as in single cell cloning.

Cereal: Edible grain such as wheat, barley, etc.

Cernuous. Pendent or drooping, as in some varieties of fuchsia, etc.

Certified stock: Plant material guaranteed to be free from virus infection and other diseases.

Cespitose: Caespitose. Forming dense mats or tufts.

Chemically defined medium: A nutrient medium for plant tissue culture in which all of the chemical components are fully known and stated. Quality (high purity) chemicals are utilized, and no undefined constituents are included.

Chemotactic movements: The movement of plant cells in response to chemical stimulation.

Chestnut compound: A mixture of copper sulphate and ammonium carbonate used to protect seedlings from fungal attack.

Chimaera: A plant whose tissues are of more than one genetic kind. This can happen due to mutations in a cell of a very young plant, or can be caused by grafting.

Chimera: Plant part consisting of tissues of diverse genetic constitution, often observed in flowers.

Choripetalous: Having separated petals.

Chromatin: An easily stained substance in the nucleus of a cell.

Chromatophore: A plastid within a pigment-cell in protoplasm or bacteria.

Chromosome: One of the microscopic bodies present in all living cells which control development and determine characteristics of the offspring both of plants and animals.

Cicatrix: Scar left where a leaf has fallen.

Ciliolate: Diminutive of ciliate-fringed with minute hairs.

Cinereous: Powdery-grey, the colour of wood ash.

Circumscissile: Opening by means of a split around the circumference so that the top comes off like a lid, as in some seed capsules. The process is also known as equatorial dehiscence.

Cirrhous: Tendril-like.

Citriculture: The cultivation of citrus fruits.

Cladode: A stem which is flattened, and thus bears a superficial resemblance to a leaf; found in some xerophytic plants, and other which may have reduced or suppressed leaves.

Classification: The act of arranging all plants in groups with a name which is understood internationally. Latin is the universally accepted language for the precise specification and naming of these groups. Botanists divide and subdivide the classes until the system is quite bewildering to the ordinary gardener, but the following is a simplified explanation. 1. The Species is the basic unit and although it has several definitions, the one which is generally accepted in the plant world is that species can readily interbreed with one another but not with members of another species. In those rare cases when they do, the hybrid offspring will be sterile. 2. A group of species (being closely related) forms a Genus. 3. A group of genera forms a Family. 4. Closely related families make an Order. 5. Related orders form a Class and above that category are Division and Kingdom.

Clonal multiplication or clonal propagation: Vegetative (asexual) propagation from a single cell or plant.

Clove: 1. One of the young bulbs of shallots or garlic developed as an offshoot of a larger bulb. 2. A type of strongly scented carnation.

Club root: A fungus disease which attacks cruciferous plants, including brassicas. The fungus lives in badly drained acid soils and may persist for many years. Prevention is by dressing the soil with lime before planting, and allowing a lapse of at least three years before replanting brassicas on the same piece of ground.

Clump: Closely compact group of roots, etc. which can be divided for purposes of vegetative propagation.

Cluster cups: Common name for burst rust producing bodies of a fungus after the spores have been dispersed. The cups resemble tiny volcanic craters left when the spores rupture the epidermis.

Co-culture: The joint culture of two or more types of cells; as a plant cell and a microorganism or two types of plant cells; as is done in various dual culture systems or the nurse culture technique.

Coalescence: Stateof growing together; in flowers, the union of parts of the same kind, as united sepals (synsepaly)

Coconut milk: or coconut water. L. liquid endosperm from the centre of the coconut seed. A complex, undefined addendum of variable quality and effects in some nutrient solutions (2-15% v/v) for plant tissue culture. It has growth promoting effects and cell division factors. It is replaceable in some cases by cytokinins and/or sugar. It was first used by van Overbeek et al. (1941) to stimulate Datura embryo cultures.

Codling: Immature apple (fruit).

Cold frame: A glazed garden frame providing protection but not heated; often used for hardening-off tender subjects before planting out.

Cold house: A greenhouse, unheated except by natural sunlight.

Cold treatment: A cold moist treatment prior to bulb planting that induces rapid shoot elongation and flowering.

Cole: Cabbage, especially the smaller varieties.

Compatible: Capable of free-cross fertilization. Fruit trees that can pollinate each other are termed compatible and varieties are selected with this in mind. Grafts also have to be compatible or the scion will not take.

Compensation point: 1. The light intensity at which the amount of carbon dioxide (CO_2) taken up by plants during photosynthesis is equivalent to that released during respiration. 2. That concentration of CO_2 at which CO_2 release from respiration and uptake during photosynthesis are equal.

Conditioning: 1. A phenotypic alteration resulting from the action of external agents during critical developmental stages. 2. An undefined interaction between media and tissues encouraging the growth of single cells or small aggregates. Conditioned medium may be used to induce growth of cells plated at low densities (below the minimum inoculums size) or cells unable to grow for unknown reasons. Conditioning may be accomplished by immersing cells or callus contained within a porous material (such as dialysis tubing) into fresh medium for a time dependent on cell density and a volume related to the amount of fresh medium.

Conservatory: A term first use in the seventeenth century to describe a greenhouse but now denoting a garden room, usually attached to the house, equipped with table and chairs in addition to plants.

Continuous culture: A cell suspension culture with a continuous influx of fresh medium, maintained a constant volume by the efflux of spent medium (closed continuous) or with the efflux or cells and spent medium (open continuous).

Controlled temperature forcing (CTF): Procedure for treating Easter lilies in which nonprecooled bulbs are potted, placed in a controlled temperature are at approximately 61°F to 45°F for 6 to 7 week prior to being placed in the green house.

Coppice: Small wooded area, mainly of small trees, grown for periodical cutting.

Coppicing: The cutting of trees back to ground level in order to promote the growth of several new stems from one root-stock.

Cordon: Usually describes a fruit tree which has been trained to a single stem, such a cordon apples, but it can also apply to other plants such as sweets peas, etc.

Corn milk: Corn endosperm tissue. An indefined complex addendum to some plant tissue culture media.

Corniculate: Bearing a small horn-like attachment.

Cotyledon: A primary leaf or seed leaf. Quite different from the true leaves which develop later, they are usually dicotyledons or monocotyledons gymnosperm seedlings, such as pine may have several cotyledons. In some plants, such as peas, the cotyledons remain underground.

Covert: Small man-made thicket originally created as a shelter for game.

Creeper: Plant that creeps along the ground or , more rarely, up a support of its own accord.

Critical daylength: The daylength above or below which a plant will flower, depending on whether the plant is a short-or long-day plant.

Crocus pot: A small fancy container having several apertures in the walls in which crocus bulbs are planted. The pots can also be used for other small plants such as aubretia, etc. larger sizes are available with more holes in the sides for planting with strawberries.

Cross fertilization: Fertilization of a female gamete of one plant by a male gamete from another.

Cross pollinate: To transfer pollen between different plants of the same species.

Crown gall: A disease of plants in which tumors form. The sausal agent is the bacterium, Agrobacterium tumefaciens. Gall tissue has been grown in vitro. A tumor inducing portion of the bacterial genome (the Ti plasmid) may be used experimentally as a genetic tool to incorporate (vector) genetic information into plant cells.

Crumb: The ideal condition of the soil when it crumbles in the hand, not too dry and not sticky. Breadcrumbs would be a reasonable comparison.

Crustaceous: Brittle, hard, gritty.

Cryoprotectant: An agent able to prevent freezing and thawing damage to cells as they are frozen or defrosted. These substances have high water solubility and low toxicity. They are classified either as permeating (glycerol and dimethyl sulfoxide) or non-permeating (sugars, dextran, ethylene glycol, polyvinyl pyrolidone and hydroxyethyl starch) agents.

Cultivar: A hybrid name derived from CULTIvated VARiety and applied to a plant that has originated and persisted under cultivation. They are usually distinguished typographically by being set in roman, not italic, type and contained within single quotation marks as in Pelargonium domesticum 'Grand Slam'.

Cybrid: A cell or plant cytoplasmic hybrid (heteroplast) with the nucleus of one and cytoplasmic organelles of another, or of both cells or plants.

Cytodifferentiation: A functional and/or morphological differentiation occurring in cells during ontogenesis, affecting their phenotype.

Cytokinin: One of a large class of plant hormones (phytohormones) or synthetic growth substances.

Cytoplasmic gynogamy: The apparent fertilization of a female gamete by the cytoplasm of the male gamete.

Cytoplasmic hybrid or cybrid: A cell or plant hybrid (heteroplast) with the nucleus of one and the cutoplasmic organelles of another or of both cells or plants.

Cytoplasmic variant: A maternally inherited change in a cellular trait.

Cytotoxic: A chemical or other agent toxic to cells.

2,4-D: 2,4-dichlorophenoxyacetic acid; a synthetic translocated hormone weed killer, use to control broad leaved herbaceous plants, and as a defoliant.

Daddock: The heart of a rotten tree.

Daisy grubber: Small hand tool having a narrow blade with a V-shaped notch at the end and cranked in the middle to provide leverage. Originally used for grubbing-out daises in lawns but now more widely use for docks, dendelions, etc.

Daminozide (succinic acid 2, 2-dimetbylbydrazlde): The active ingredient in the commercial products. B-Nine and Alar (SADH). Daminozide(suceinic acid 2, 2-dimethylhydrazide). The active ingredient in the commercial products B-Nine and Alar (SADH).

Damping down: The spraying of paths and benches in a greenhouse with clean water in hot weather to reduce the temperature and improve the relative humidity of the atmosphere.

Dark storage: A term used to describe the time foliage plants remain in darkness within shipping containers during transit or storage.

Daughter bulb: Scales and leaves initiated by and developing below and around the new daughter apex. This apex arises from a bud in the axil of a scale subtending the old or mother axis.

Day-length: The length of daylight hours has a controlling effect upon the initiation of the flowerbuds of many plants. Those which flower when days are long are called 'Long-day plants', and those flowering when daylight is shorter are termed 'short-day plants'. 'Day-neutral' plants are not affected by the day-length and are called photoperiodismic. Both long-day and short-day plants are encouraged commercially to flower outside their normal season by manipulating the illumination of the subjects to produce artificial lengthening or shortening of the daylight hours.

De novo [Latin: from the beginning, a new]**:** Arising, sometimes spontaneously, from unknown or very simple precursors.

Dead-heading: The removal of dead flower heads from plants, essential if further flowers are desired. Dead-heading prevents the development of the seed heads, quite apart from keeping the garden tidy, for if seed is allowed to develop, the production of further flowers will cease or certainly be reduced it is especially important on annuals which normally die after producing seed.

Deblossomlng: The removal, either partially or totally; of blossom from top fruit trees. Total deblossomings is usually applied to fruit trees in their first season after planting to enable the tree to establish itself before fruit production begins, if a tree, is allowed to fruit immediately after it is planted, its growth will be inhibited and it will be several years before it recovers and gets into full production. Partial deblossoming is practiced when there is an excessive amount of blossom, which, if allowed to develop into fruit, would seriously strain the tree, with a possible risk of tree clevelopingthe habit of biannual bearing.

Decalcify: Deprive of lime or calcareous matter.

Decomposttion: Metabolic degradation of organic matter into simple organic and inorganic compounds, with consequent liberation of energy; decomposer.

Differentiation: The resumption of meristematic activity by more or less mature cells through a reversal of the process of cells or tissue differentiation. Cell division leading to the formation of small, microvacuolate, isodiametric cells with prominent nuclei, that are capable of organized development; as when adventitious buds or roots develop on mature tissue.

Deficiency disease: A disease resulting from a diet that lacks an essential nutrient, such as a vitamin, essential amino acid or trace element.

Deficiency: 1. An insufficient supply or unusable form, or one or more nutritional or environmental requirements, for plant development, growth or physiological functions. The absence of adequate conditions for plant growth or performance, resulting in disease. 2. The deletion of a gene or a series or genes. This may result in a mutant phenotype or may be lethal.

Defloration: The act of stripping off flowers of removing flowerheads; usually applied to the gathering of healthy blooms as opposed to dead heading-removal of dead flowerheads.

Defoliant: Any substance or mixture of substances for which the primary use is to cause the leaves or foliage to fall from a plant.

Deforestation: Artificial clearance of all or part of a forest area. Total deforestation rarely occurs naturally.

Defruiting: The removal of all immature fruit from a tree to allow it to build up its strength. Usually done when deblossoming has been overlooked or forgotten at the correct time.

Dehiscence: The opening of a plant organ when mature, to shed its contents, e.g., anthers, some fruits, moss capsules, etc.

Dehiscent: Applied to fruits which split open to release their seeds when ripe, such as peas, lupins, etc., also to anthers which split to release their ripe pollen.

Dehorning: A severe method of pruning elderly fruit trees to bring them back into production or to reduce the overall size. Some of the larger branches are totally removed, and all dead and decaying wood cut away, together with any inward-growing branches, to open up the centre and permit more air movement and light penetration. The remaining branches are often cut back to about half of their length. All cuts should be carefully trimmed and covered with a protective coating.

Delphinidin: 3, 3, 4, 5, 5, 7 hexahydroxyflavylium cation. The aglycon of many Anthocyanins. Their glycosides are widespread among plants and are responsible for the mauve and blue colours of many flowers and fruits.

Dendrology: The study of trees and the natural history of trees.

Dendrophilous: Thriving in trees; living in orchards; dendrophile, dendrophily.

Dene: Steep-sided, heavily wooded small valley

Dense: Crowded together, compacted thick.

Dentate: Coarsely toothed, with the indentations rather spreading and regularly divided, as applied to a leaf margin.

Denticulate: Diminutive of dentate-having smaller teeth.

Dentition: The arrangement of teeth at the edges of leaves, and occasionally applied to some petals.

Depauperate: Starved; poorly developed or distorted due to unfavourable conditions.

Depressed: 1. Sunken, as in some necrotic spots on foliage. 2. Flattened

Detergent: A cleaning agent; one of numerous synthetic cleaning preparations chemically different from soap. Detergents are commonly used to prepare work areas for aseptic plant tissue culture manipulations and to remove dirt and microorganisms from plant material prior to explantation.

Devernalization: A process where stimulus to flowering received by, treatment with low temperature (i.e., vernalization) can be reversed.

Dewar flask: A double-walled. glass flask used to keep liquids at other than ambient temperatures; as a thermos does.

Dewpoint: Air temperature at which dew begins to be deposited. After a warm day m the garden the temperature fails, the air becomes so humid that it is saturated and condensation is visible on smooth surfaces. The foliage will usually, show dew first because the transpiration, process cools the leaf surfaces and thus lowers the temperature.

Diadelphous: Having stamens which are united by the filaments in two bundles as in peas and beans.

Diagcotropic: Transversely geotropic, said of rootsor leaves which grow horizontally.

Diageotropism: Development of plant part by growth curvature in response to stimulus of gravity, e.g., geotropism.

Dibber: Instrument used for making a hole in the soil into which a seed, seedling, cutting or tuber can be inserted. Garden shops sell specially made tools, but anything can be used, such as a pencil, garden cane or fork handle, according to the diameter of hold required. To dibble is to use a dibber.

Dichlamydeous: Of flowers possessing a double, perianth. i.e., calyx: and corolla.

Dichotomous: Divided repeatedly and regularly in pairs with, the two branches of each fork roughly equal.

Dicotyledons: Angiosperms having two embryonic first leaves or cotyledons; leaves usually have netvenation, and flower parts are in whorls of two, four or five.

Die-back: Any of several fungus diseases which, cause shoots or branches usually to die back from the growing tips. It is a much slower process than wilt, where infected shoots collapse rapidly and entirely. Diseased shoots should, be cut back to sound wood and the infected material burned.

Dientomophilous: Used of a plant pollinated by two different insect species and having two kinds of flowers each adapted for one of the insect, pollinators; dientomophily.

Digging hoe: The digging or heavy hoe is mainly used for heavy work on stony ground. It is used in much the same way as a pick and has been a basic agricultural implement for more than 2000 years.

Dilated: Roughly flattened and expanded into a blade.'

Dimerous: Applied to flowers or whorls of a flower, where each whorl consists of two parts.

Dimethyl sulfoxide [DMSO, C_2H_6OS mw 78:13): A highly hygroscopic liquid with little odour- or colour. It is an organic cosolvent sometimes used in small quantities to dissolve neutral organic substances in plant tissue culture media preparation. Readily penetrates skin. An extremely powerful solvent. It can usually be replaced by a less toxic solvent It also has uses as a cryoprotectant.

Dimorphic: Two different forms in the same species, as in the adult and juvenile leaves of ivy.

Dimple: Old term used to describe the virus disease of pears, now, called stony pit. The fruits show small dimples or depressions on the skin, below which are hard gritty lumps.

Dingle: Peep hollow surrounded by or shaded with trees; a dell.

Diplococcus: Type of micro-organism which forms pairs.

Diplostemonous: Applied to flowers which have stamens in two alternating whorls, the outer whorl alternating with the petals.

Dipterous: Having a wing-like appendage, as of certain seeds.

Direct embryo genesis: Embryoid formation directly on the surface of zygotic or somatic embryos or on seedling plant tissues in culture, without an intervening callus phase.

Direct organogenesis: Organ formation directly on the surface of relatively large intact explants, without an intervening callus phase.

Disbudding: The removal of lateral flower buds on stems of plant, such as the carnation and chrysanthemum.

Disbudding: The removal of side shoots and surplus buds to force the plant to concentrate its energies in the remaining single bud. Exhibition carnations, dahlias, chrysanthemum, etc.

Disc florets: Florets in center of chrysanthemum inflorescence, conspicuous in daisy-or anemone-flowered types.

Disc flower: One of the central, symmetrical flowers of the capitulum of composites.

Disc: 1. The centre part of a flower such as the daisy, where the outer florets are attached to a central cushion or disc which often secrets nectar. 2. The flattened part at the end of a tendril used for adhesion.

Discoid, Discodal: Disc-shaped; circular and flattened.

Discolorous: Not of the same colour. The upper and lower surfaces of a leaf are often discolorous.

Disease: Almost any ailment of plants other than pest damage. Attacks by bacteria, fungi, viruses and sometimes the results of poor growing conditions, are all grouped under the general term disease.

Disiccant: Any substance or mixture of substances which when applied to plant, causes dehydration of the tissue.

Disinfect: To free from infection by destroying disease-causing micro-organisms, usually by means of a physical or chemical agent (disinfectant) which kills infectious organisms. This process is disinfection.

Disk: 1. A flat, circular plate-like or curved spore producing part or a fruit body in Discomycetes. 2. In a pileus of an agarics, the central part of the top surface. 3. In a flower a flat, round receptacle.

Disorder: Ailment caused by poor conditions, mineral deficiencies or a hostile environment as distinct from one emanating from attack by pests.

Disorders (Physiological): Undesirable effects caused by a nonpathogenic agent or factor.

Disperse: To scatter or distribute seed in different directions. Dispersal can be by many different agencies, including wind, water, animals, etc.

Displanting: The word used for transplanting during the early eighteenth century.

Dissect: To cut, expose, separate or divide the parts of a plant or animal into sections.

Dissected: Deeply cut into narrow segments, as of certain leaves.

Disseminule: A disseminated propagule; a seed, fruit spore or other. structure modified for dispersal.

Dissepiment: Partition or dividing membrane.

Distichous: Having two opposite, vertical rows, as applied to leaves or branches.

Distinct: Separate, not united (as in petals).

Distorted: Growing into unusual or crooked shapes, as in petals, leaves, stems, etc. Normally the result of viral or insect damage.

Distributor: Mechanical device for spreading fertilizer and similar materials evenly. Small types can be used on garden lawns, larger ones are towed behind tractors for farm use.

Diurnal: Belonging to the day-time. Usually applied to flowers which open only during daylight hours, e.g. dandelion.

Divaricate: Widely divergent.

Divergent: Spreading but to a less extent than divaricate.

Divided: Separated but not completely detached from the midrib of a leaf.

DNA: Deoxyribosenucleic acid, a molecule found in chromosomes of animals and plants, now known to contain the coded information carried in the genes responsible, for the development of individual character.

Dome: Canopy of trees, vines, etc. formed by the foliage.

Donor plant (mother plant): The source that plant used for propagation, whether an explants, graft or cutting. Source plants used for micropropagation are usually poathogen-tested (disease-free).

Dormin: Name of hormone stimulating dormancy.

Dot plant: Plant used in bedding schemes to provide height and contrast. Usually among shorter varieties or ground cover plants. Standard roses, dwarf trees, standard fuchsias and pelargoniums are among those frequently used as dot plants.

Double leader: A tree which produces twin shoots at the apex instead of the more normal single spike.

Double: Flowers which have the normal complement of petals are termed single and those with more than the normal are called double. A flower with only slightly more than the single amount is termed semi-double although to be precise, this is a contradiction in terms. Many fully double flowers are sterile because the extra petals have replaced the stamens or pistils.

Doubly serrate: Of leaf margins in which the primary serrations bear smaller, secondary teeth.

Downs, Down land: Hilly, treeless open land, widespread in the south of England.

Downy: Usually applied to leaves which are covered with short stiff hairs.

Draw hoe: Long-handled tool with metal blade for taking out seed drills, etc., it is pulled towards the user.

Drawn: Etiolated. Plants grown in poor light or crowded conditions will become tall and spindly, and will often lose much of their green colouring. They can never fully recover from this condition, and the mature plant which develops from a drawn seedling will almost always be inferior in yield, whether of fruit or flower. Drawn plants are often described as 'leggy'.

Drench: 1. To wet thoroughly; to soak. 2. To apply liquid to the soil rather than to the foliage of a plant.

Drepanium: A cymose inflorescence with branches on one side only; and often curved to one side as in some members of the rush family.

Dressing: Any fertilizer, whether organic or chemical, applied to the soil in a dry as opposed to a liquid from. Base Dressings are incorporated into the soil before planting; top dressings are applied to the surface around growing plants. Top dressings of bulky materials such as animal manure are usually called mulches.

Dribble bar: A perforated tube of metal or plastic, sometimes called a sprinkle bar, which can be attached to a watering can for the precise application of fertilizers. Weed-killers, etc. Wide forms are available for use on lawns, narrow ones for better control around plants

Drill hoe: Normally a long-handled hoe with triangular blade for taking out seed drills, but there are also short-handled versions.

Drum, Drumhead: Flat-toped variety of cabbage.

Drupe: A stone fruit where the seed is protected by the hard casing (stone) contained in a fleshy layer within an outer skin. Peach and apricot are simple drupes; raspberry and blackberry are compound drupes.

Drupel: A small drupe, usually in collections, e.g., raspberry.

Drupelet, Drupel: A small drupe or the individual of component of a compound drupe, as in blackberry.

Dry rot: A loose term covering several diseases in plants such as bulbs, corm, tubers and fruits. Symptoms may include brown patches, shrivelling areas of dry decay, etc. Infected items should be burned.

Dry set: Applied to glasshouse-grown tomatoes which fail to grow larger than pinhead size, usually as a result of an overheated dry atmosphere. There is no cure but the problem can be prevented by regular overhead spraying with clean water or a fruit-setting agent.

Dry weight: The moisture-free weight of tissue obtained by drying at high [oven-drying] or low [freeze drying] temperatures for an interval sufficient to remove all waters.

Dry-pack storage: The storage of cut flowers in vapor-proof containers with the stems not in water, usually at 32°F

Dual culture: A culture system that includes plant tissue and one organism [such as a nematode species] or micro-organism [such as a fungus]. Dual cultures are used to study host-parasite interactions or in the production of axenic cultures for a variety of purposes. Usually the microorganism selected is an obligate parasite.

Duct: Part of a plant's vascular tissue carrying air, water, etc.

Duramen: The dense centre of a tree trunks, the hearwood.

Durian: Durian (Durio zibethinus) is one of the best known fruits in Asia, where it is, seen as a gourmet's delicacy and referred to as the 'King of Fruits. The fruit is known for its strong and pungent odour, which can be offensive and nauseating to some people. It is widely cultivated throughout Southeast Asia and is one of the most important economic fruit crops of, Thailand, Malaysia and Indonesia. Durian has been grown in these countries for several hundred years. It is one of the fruits with the highest export value. Besides selling as a fresh fruit, processed durian is gaining an increasing share of the market.

Durio gravseolens: This species has the greatest diversity and is extremely distributed. The pulp of D. graveolens has a most exciting spectrum of colours, ranging from tantalizing crimson red to pale yellow, evsen whitish. The taste varies, and the fragrance can be impressive and is reputed to leave a long lingering aftertaste.

Dusting: The application of dry powdered fungicides, pesticides, etc. through a mechanical device designed to produce a fine, even fog. Also known as fogging.

Dutch hoe: Long-handled tool with metal blade for loosening soil, weeding, etc. it is pushed away from the user.

Dutch light: A single sheet of glass about 59 x 31 in (150 x 79 cm) enclosed in slots in a wooden edging without putty. They can be used in the construction of garden frames, or bolted together to make greenhouses.

Dwarf male: 1. Of bryophytes, miniature antheridial gametophores that are attached to the normal-size archegonium bearing gametophores. 2. Of algae, small male filaments of nannandrous species of Oedogonium developed by the germination of andiospore.

Dwarf: Extremely small. Applied to plants that are naturally tiny; plants which have been artificially dwarfed by root restriction or grafting, and plants which have been chemically dwarfed for one season, after which they revert to their natural form.

Dwarfism: The conditions of being stunted, much smaller than normal; having restricted growth; micro somia; nanism; dwarf.

Ear: Head of corn or ripe com-spike containing seeds.

Earthworm: Beneficial terrestrial animal which lives normally in the soil (although some species can survive in water), mainly nocturnal, spending day light hours underground and coming to the surface to feed. As a general rule, the more worms to the acre, the more fertile the land. They can be a nuisance to pot plants growing in unsterilized soil.

Earwig: An insect related to grasshoppers, cockroaches, etc. having a long; fairly narrow body with a pair of nippers or pincers at the rear. Basically seavengers, earwigs can cause severe damage to some plants, especially those with a dense flower head such as dahlias, which provide food and shelter. Mainly nocturnal feeders, they can be trapped by placing paper or cardboard dunces hats stuffed with dry straw among the flower heads, inspecting, them each morning and destroying the occupants.

Edging tools: A variety of tools used to trim and maintain the edges of lawns such as straight-bladed types of spade, right-angled long-handled shears, electric trimmers, etc.

Edging: 1. The use of materials such as wood, metal or stone to keep lawn edges and paths adjacent to borders tidy. 2. Using small, compact plants to outline flowerbeds. In the past box and dwarf lavender were favorite subjects for this purpose. 3. Trimming lawn edges to give a sharp, clean-cut Finish after mowing.

Efflorescence: The bursting or opening of the flower bud; breaking into flower.

Eglandular: Lacking glands.

Elaioplast or lipidoplast: An oil-storing plant cell plastid. A specialized leucoplast.

Elaiosome: Of seeds which have or contain an oily substance to which ants are attracted.

Electronic leaf: Part of a mist-propagation system which controls the supply of liquid to the spray head.

Endomembrane system: A system of cytoplasmic membranes found in eukaryotic cells. It participates in the synthesis, transport and storage of chemical compounds within the cell and facilitates export from cells. It includes the plasma membrane and tonoplast of vacuoles.

Entomophilous: Pollinated by insects.

Entomophily: Pollination by insects. Flowers pollinated by insects are usually brightly coloured and scented. If they are pollinated by bees, they usually produce large amounts of pollen which the bees collect. If they are pollinated by butterflies or moths they produce nectar; entomophilous.

Environmental chamber: A controlled environment cabinet [incubator] in which temperature, light quality, intensity and duration; and preferably also the relative humidity and airflow are controlled.

Epichil, Epichile: The termination of the lip in some orchids when distinct from the basal part.

Epicotyl: Seedling shoot above the seed-leaves; the emerging growing-point.

Epipetalous: Having stamens arising from or carried on the petals or corolla.

Epiphyte: A non-parasitic plant which uses another as its host, often having aerial roots fur the collection of atmospheric moisture. Some orchids and many bromeliads are epiphytic.

Epithet: The second part the specific epithet of the binomial or the third part the sub specific epithet of the trinomial.

Epizoic: Having seeds which rely upon animals for dispersal, such as goose grass which produces seeds with hooked prickles.

Epsom salt: The common name for magnesium sulphate, a vital element in the production of chlorophyll in a plant. A deficiency shows in the discolouration of the leaves between the veins, developing into dead areas if the condition is allowed to persist. The remedy is to spray is to spray the foliage with a foliar feed containing magnesium sulphate.

Equitant: Overlapping so as to form a type of fan, applicable to leaves, e.g. the iris.

Established culture: 1. Achievement of an aseptic viable explant, synonymous with Stage I culture [culture establishment] or transplant, synonymous with Stage IV culture [establishment in soil]. 2. A suspension culture subjected to several passages, adjusted so that cell number per unit time is constant from subculture to subculture.

Ethephon: The active ingredient in the commercial products Ethrel and Floral, which is metabolized to ethylene, in the plant. Ethephon or 2-chloroethyl) phosphoric acid [$ClC_2PO_3H_6$ mw

144.50]. A synthetic compound .that spontaneously degrades to release ethylene. Often used as a means to treat cultured cells [or un ripened fruit] with ethylene.

Ethylene: Colorless, odorless gas that hastens senescence of flowers. Often emitted by fruit, foliage, and incomplete combustion of oil and gas in heaters.

Ethylene scrubbers: Chemicals used to remove ethylene from the atmosphere.

Even-pinnate: Applied to a leaf composed of leaflets in pairs but without a terminal leaflet. When a terminal leaflet is present the leaf is described as 'odd-pinnate'.

Evergreen: A plant having leaves that remain green and functional throughout the year.

Ex vitro: [*Latin; from glass*]. Organisms removed from culture and transplanted, generally to soil or potting mixture.

Explant donor: The source plant or mother plant from which the explant used to initiate a culture is taken.

Explant: The excised plant portion used to initiate a tissue culture. The process of dissection and removal of culture of these small organs or tissue sections is explanation. Explants choice, the timing of excision and pretreatment are important determinants of culture success.

Explosive fruit: One which bursts suddenly and violently, scattering seeds over a considerable area, as in the cucumber.

Exponential phase: A phase in culture in which cells undergo their maximum rate of cell division. It follows the lag phase and precedes the linear growth phase in most batch propagated suspension cultures.

Exserted: Protruding or projecting organs such as stamens or pistils which protrude beyond the rest of the flower.

Exsiccata ***pl.*** **Exsiccatae:** A dried herbarium specimen.

Exstipulate: Lacking stipules, as in privet.

Extra-foral: Outside the flower, as of nectaries on leaves.

Extrorse: Applicable to anthers that turn and shed their pollen outward, opening toward the outside of the flower.

Exude: To discharge slowly, leak liquid material [exudatel sometimes through pores or cuts but often by diffusion into the medium. This process of exudation is associated with a lethal browning of explants in some woody plant species. These in vitro exudates have not been chemically characterized but are often referred to as tenins or oxidized polyphenols.

Eye: Usually an undeveloped growth bud, such as the eye of a potato: but can also refer to the centre of a flower, as in some primroses where the eye is a different colour or shade to the rest of the flower.

F_1 Hybrid: First-generation cross between two pure bred strains. Seed is produced by controlled hand pollination but the second generation will seldom breed true. The crop is usually very consistent in size, colour, vigour. etc.

F_2 Hybrid: Second-generation cross between two F_1 hybrids. Not so vigourous as F_1 hybrids and usually showing great variation.

F_2: Second generation following hybridization.

Factor: 1. A condition, influence or constituent to be taken into account. 2. A genetic character determinant (gene).

Facultative parthenogenesis: The process by which some eggs develop parthenogenetically if not fertilized.

Fan Palm: The European fan palm has palmate rather than pinnate leaves, and is the only palm indigenous to Europe. It is not really hardy in/the British climate but will withstand temperatures almost down to freezing point. The Canary Island palm is indigenous to the Canary Islands politically but not geographically in Europe.

Fanaesol: A colourless liquid extracted from oils of plant such as citronella, neroli, cyclamen, and tuberose. It has delicate floral odour.

Female flower: A flower that has no stamens but contains functional carpels. Such a, flower can only reproduce after fertilization by pollen from a male flower of the same species.

Fermentation: The anaerobic energy yielding breakdown [decomposition] of complex organic substances, especially carbohydrates like glucose, by microorganisms. It is often misused by tissue culturists to describe large scale aerobic plant cell culture for secondary product synthesis.

Flabellate: Fan-shaped, or a fan like structure. The foliage of the Chusan palm is flabellate.

Flocculation: A method of improving soil texture by the use of lime or others soil conditioners, thereby increasing the air content, especially in. heavy clay soils.

Flora: The complete range of plants growing in particular area or climate, or a book that catalogues these.

Floral primordium: Very early stage of flower bud.

Floral: Anything pertaining to flowers- petals, sepals, anthers, etc.

Floricane: The biennial cane of bramble, raspberry, etc. during its second year, upon which flowers and fruit are produced.

Floriculture: Victorian expression for the cultivation of flowers.

Florigin: A hypothetical plant growth substance supportably associated with the onset or regulation of flowering.

Flower: A specialized shoot of a plant bearing the reproductive organs, often brightly coloured to attract the fertilizing insects; the blossom of a plant.

Flowerbed: A garden bed for flowers; often a section of open ground cut into a lawn or paved area.

Flowering hormones: Hormones which initiate the formulation of floral primordia, or promote their development.

Flowering plant: A plant which bears flowers, This term is usually used only for angiosperms but is sometimes used for some gymnosperms as well.

Flowers of sulphur: Refined sulphur dust used mainly under glass against mildews, but also to control red spider and other mites.

Flue dust: Soot formerly used as a nutritional additive. The sweepings from modem high temperature boilers have little food value for plants; it bears no comparison with old-fashioned soot from coal-fired boilers or the chimney stacks serving coal fires.

Forestry: The science of forest management; the art of planting, tending and managing forests.

Fork hoe: A type of hand-cultivator with two or three prongs instead of the more usual blade.

Formal garden: A garden laid out in geometrical designs such as square and circular flower beds, often including clipped hedges and architectural features, fountains, etc.

Foveolate: Pitted, having one or more small depressions. Bitter pit of apples was once thought to be the result of virus attack but is now confirmed as emanating from poor growing conditions.

Frill-girdling: A method of destroying established trees by making downward-sloping, overlapping cuts with a hatchet or similar implement, encircling the trunk at or just above ground level, then filling the cuts with a powerful brushwood killer.

Frost Pocket: Small area subject to frost. If cold air, as it flows downhill, meets an obstacle such as a wall a frost pocket may form where the air temperature is considerably lower than it is above, presenting a hazard to tender plants.

Frost resistance: Varying ability of certain plants to withstand frosts. Plants which are totally frost resistant are termed hardy and those which are not, tender; half-hardy and near tender convey some degree of frost tolerance.

Fructiferous: Fructuous. Bearing fruit; fruitful.

Fructification: 1. The process of producing fruit. 2. A fruit and its appendages. 3. A general term for the body which develops after fertilization, and contains spores or seeds. 4. Any spore producing structure, whether developed after fertilization or vegetatively, as in fungi an ascocarp, basidiocarp or an ascervulus.

Fruit gatherer: A mechanical device for gathering fruit from tall trees without damage, consisting of a pair of clippers on. a long pole with a bag or similar receptacle fixed just below the clipper head to catch the fruit. There are several variations to the basic design to suit the particular fruit to be harvested.

Fruiting body: The spore-bearing organs of the larger fungi, including mushrooms, toadstools, common puffball, death cap.(fatal if eaten), etc.

Fruitless: Failing to achieve the desired results, unproductive.

Funicle: The stem of an ovule by which it is attached to the placenta in the ovary. Also known as seed stalk.

Genus: Group of closely related plants, forming a subdivision of a family, itself further subdivided into species; the first word of the Latin or scientific name denotes the plant's genus.

Germination: Sprouting of a seed to produce a new plant.

Gibberelic acid (GA): Chemical compound used to break flower bud dormancy or stimulate shoot elongation.

Green pruning: Pruning of actively growing rose plants without benefit of a dormancy period.

Greenhouse phase: That portion of forcing that encompasses the time from placing the plants in the greenhouse until flowering.

Gynandrium: The structure in the orchid flower that results from the fusion of the male and female portions of the flower.

Gall: 1. An abnormal growth on a plant caused by the parasitic attack of fungi, insects or bacteria e. g., the oak apple. 2. Bare spot in a coppice or field.

Gametoclone: A plant regenerated from a tissue culture originating from gametic tissue.

Gametophyte: A plant producing gametes, the stage (of a plant having alternating generations) which produces gametes, e.g., the prothallus of a fern.

Gel: A mixture of thick or firm consistency containing liquid trapped in a solid component; a gelling agents such a agar solidify liquid nutrient media. The process of gelation.

Genetic engineering or recombinant DNA technology: Technology involving man-made changes in the genetic constitution of cells [apart from selective breeding]. This technology usually employs a vector [such as the Ti plasmid of Agrobacterium tumefaciens] for transferring useful genetic information from a donor organism into a cell organism that does not possess it. Genetic engineering has many potential uses.

Genetic selection: Selection of genes, cells [cell selection], clones, etc., by man within populations or between populations or species. The usual purpose is to alter a specific phenotypic character.

Such selection usually results in differential success rates of the various genotypes, reflecting many variables including selection pressure and genetic variability in populations.

Genetic variation: Differences in individuals derived from the same genotype in distinction to difference caused by the environment. Genetic variance denotes the proportion of phenotypic variance caused by differences in the genetic make-up of an individual.

Geophytes: A plant growing from underground organs such as tubers.

Gibberellins antagonists: Inhibitors of the effects of gibberellins on plants. Their effects can be at least partially overcome by gibberellins. The term is used independently of the mechanism of inhibition. Competitive inhibitors of gibberellins, compounds which bind to the same active sites as the hormones are called antigibberllins.

Goblet: A method of training a fruit tree in the shape of a goblet or vase. The branches arising from the short trunk are trained outward at first, then upward to create a cup-shaped framework. A wooden or metal hoop added as they grow to maintain the shape. Said to result in increased fruit production, but in any case makes a very attractive and unusual ornamental feature.

Grafting: The artificial union of one part of a plant with parts of another (stock), e.g. when one variety of apple is grafted on to a different rootstock.

Grip trench: A slit trench made by inserting a spade vertically into the soil and rocking it slightly to produce in slit in which cuttings can be set.

Growth rate: A measure of the rate of growth of an organism of culture.

Growth retardant: A natural or synthetic chemical substance that inhibits plant growth.

Growth water: That portion of the total soil water available to plants chresard.

Guttation: Droplets of water that appear at the tips and margins of leaves, usually on greenhouse or indoor plants. It is usually a sign that the compost is being watered excessively.

Gymnosperms: Flowering plants whose ovules are borne exposed on the carpel, and not enclosed in an ovary, e.g. a conifer.

Gynaecandrous: Having staminate and pistillate flowers on the same spike.

Gynaeceum, Gynoecium: The female parts of a flower.

Hack. 1. To cut or notch. 2. Mattock or type of pick used for working heavy or stony land.

Half-hardy: 1. A plant which can be grown outdoors in the summer months but which will not survive frost, e.g. runner beans, tagetes, etc. 2. Shrubs and herbaceous subjects that will survive an average winter outdoors but only in sheltered positions; or in favoured areas of the country such as the western counties of England and parts of the western isles of Scotland.

Halophyte: A plant which will tolerate an abnormal amount of salt in the Soil. Plants which grow on the seashore and sand dunes are halophytes, such as the common salt-marsh grass (*Puccinellia/ maritima*).

Hanging basket: A plant container designed for hanging in a porch, under eaves, on walls, in greenhouses, etc. They are manufactured from galvanized wire, wood, cane, plastic, pottery and wrought iron in a variety of shapes, patterns and designs.

Haploid: Germ cell having half the normal number of chromosomes.

Heart-rot: Decay of the internal plant tissues often seen in root vegetables that appear to be quite sound until cut open; often caused initially by slug damage which becomes infected by bacteria. Heart-rot in trees is the result of fungus attack.

Heart shaped: Cordate; of leaves that are pointed at one end and have a pair of rounded lobes at the other.

Henna: *Lawsoma inermis.* An old world plant having small opposite leaves and axillary panicles of which flowers; a reddish-brown dye extracted from the leaves is used in hair dyes; Egyptian henna.

Herbaceous border: Garden border stocked usually with perennial flowering plants.

Herbaceous perennial: A herbaceous plant with perennial underground parts which develop aerial shoots each year.

Herbaceous: Of plants with soft top growth rather than woody growth; they can be annual. Biennial or perennial, but in most cases the top growth dies back in the winter. Often applied to the taller non-woody perennial varieties such as golden rod.

Herbal: Book containing lists and descriptions of herbs used for medicinal or culinary purposes.

Herbarium: 1. A plant museum, or any place or building used for housing a classified collection of preserved plants. 2. A collection of plants, usually dried, contained together in a show-case or room.

Herbicide: A chemical agent that kills or inhibits the growth of herbaceous plants especially weeds.

Hermaphrodite: Bisexual flowers which have both male and female organs, so do not require a separate pollinator unless the plant is self-sterile.

Hill-up: Almost obsolete term; to bank the soil around the stems, as in the ridging of potatoes.

Hillock: Small mound or heap of soil, e.g. a molehill.

Hilt: Handle of a tool, especially a small hand-tool such as a trowel.

Hilum: The scar or mark on a seed where it was attached.

Hip: The fruit of the rose, in particular the 'dog rose' (*Rosa canina*): The botanical name for it hip is *cynarrhodion.*

Homozygous: Where both genes of a gene-pair are identical. Plants will breed true if most of their genes are homozygous. A self-pollinated plant will not produce entirely homozygous offspring unless it is entirely homozygous itself.

Honey-guide: Markings on the perianth, usually in the form of spots or lines, said to direct insects to the nectar in a flower.

Honeydew: The sticky substance excreted by sapsucking insects such as aphids and scale insects, creating a tacky coating on leaves and stems. Aphids are farmed by ants to yield a supply of honeydew as a diet supplement.

Hop manure: Spent hops from breweries, useful for improving the texture of soils and adding humus. On its own has no great food value but various chemicals are normally added to improve its nutritional functions.

Hortal: Used of an ornamental plant that has escaped from cultivation and can be found growing wild.

Hump: Small rounded hillock or boss of earth.

Humus: The dark brown or black substance produced by the slow decomposition of organic matter in which the bacteria multiply, breaking down the various components into the simpler chemicals which plants absorb as food.

Husk: The outer covering of some fruits; the shell; the dry outer case; as of corn and many other cereals.

Hynopsis: A state of dormancy in seeds that retain the capacity for normal development ; hypnotic.

Hypobaric store: Storage at less than atmospheric pressure for long-term holding of plant material.

Ichneumon flies: Loosely used term covering a large number of insects whose larvae prey upon other insects, and are therefore beneficial to gardeners. The female of Ichneumon suspiciosus, for instance, lays her eggs inside the bodies of various caterpillars in springtime, and the grubs develop within the host, devouring its body as they mature.

Immunology: The study of immunity from disease.

Impermeable: Not permitting the passage of fluids.

Implant: An insert, e.g. a scion graft.

Implantion: The insertion of a graft into the host or stock.

In situ: In the natural, original place or position: as in the location of the explant on the mother plant prior to excision.

In vitro: [Latin in glass]. Experimentation on organisms or portions thereof in glassware or culture; growing under artificial conditions as in tissue culture. The antonym of in vivo.

Incubator: An apparatus providing controlled environmental conditions, [light, photoperiod, temperature, humidity, etc.] suitable for [incubating] plants or plant cultures. This term is sometimes used as synonym for cabinet [growth cabinet] but usually implies a greater degree of environmental control.

Incurved Incurving: Bent inward, as applied to a type of florist's chrysanthemum in which the petals turn loosely inward and upward to produce a compact spherical flower.

Indexed plants: Plants that have been tested by pathological methods and found to be free of known pathogens. Plants may also be indexed for one specific pathogen.

Indicator plant: A plant showing a reaction [symptoms] to an abiotic or biotic [pathogenic] factor used to assist the detection and identification of an environmental factor or pathogen,

Indoleacetic acid (IAA) $C_{10}H_9O_2N$: A decomposition product of tryptophan produced by bacteria and occurring in urine and faeces, used as a hormone to promote plant growth. This is the most common auxin produced mainly in meristmatic tissues such as those of young developing leaves, flowers and fruits.

Infertile: Not fertile; applicable to soils deficient in nutrients or to fruit trees and bushes which fail to crop.

Inflorescence: The flowering part of a plant; individual florets or a group comprising a flower head. A general imprecise term used to describe the arrangement of the flowers on a stem or branch; the mode of flowering. Inflorescences can be of several types.

Infructescence: A cluster of fruits resulting from an inflorescence.

Insect: A small animal with body divided into segments. Over a million have been listed so far world-wide and new ones are being discovered every year. Some are harmful, others beneficial: some have wings, others are wingless.

Insecticide: A chemical used to kill insects; used generally for any chemical agent used to destroy invertebrate pest.

Instant gardening: Producing an instant effect by planting mature, often flowering subjects purchased from nurseries or garden centres in containers. In this way the seasonal planting restrictions which apply to many plants can be largely overcome.

Intercropping. 1. Growing two crops in succession on the same piece of ground. 2. Raising a quick growing crop between rows of slower growing plants. 3. Planting bush fruits in an orchard between top fruit trees, e.g. gooseberries between apples.

Isobilateral: With corresponding opposite parts alike, applicable to leaves where both surfaces are identical; as in iris.

lsodiametric: Having equal diameters. Used to describe plant cells.

Jardiniere: Ornamental stand or container for displaying a growing plant, normally used indoors.

Jenneting: A type of early apple, now seldom grown, certainly not commercially.

Jerusalem antichoke: *Helianthus* tuberosus. Compositae. A hardy perennial; tubers appear like potatoes and are cooked, pickled, or eaten raw. It is a good source of a carbohydrate inulin, a good food for diabetics.

Juice: The liquid content of fruits or vegetables.

Kainite: A natural mineral found in saltbdeposits, seldom used today except occasionally as a winter dressing for orchards. It was originally the K in NPK formulations which encouraged flower and fruit development, containing sulphate of potash, sodium chloride, sulphate of magnesium, etc. Modem formulations of high potassium fertilizers are much purer and their effects more predictable.

Keiki paste: A hormone compound sometimes applied (not always successfully) to an orchid mode in an effort to alter the cell structure so that a keiki will be produced instead of a flower.

Kernel: A seed within a hard shell; the edible part of a nut.

Kex: A dry, usually hollow, herbaceous stem, or any tall umbelliferous plant.

Key: A samara; winged seed as produced by sycamores and maples.

Kidney potato: Any long oval-shaped variety of potato tuber, e.g. the 'Jersey Royal'.

Kinetin $C_{10}H_9ON_5$: A cytokinin formed in many plants which has a stimulating effect on cell division.

Kitchen garden: A functional rather than ornamental garden reserved for the production of vegetables and fruits for culinary use; often surrounded by a wall in earlier times, and including greenhouses for raising out-of-season produce.

Knife: There are several types of knife specially designed for garden use such as the budding knife, grafting knife, pruning knife, dissecting knife, etc.

Knitch: The colloquial name for a faggot in some parts of the country.

Knoll: Hillock; small hill or the top of a hill.

Knop: Obsolete word for a knob, tuft or bud.

Knot garden: Flowerbeds or, more often, low hedges laid out between paths in such a manner as to form intricate patterns when seen from above. Early examples -in the sixteenth century were constructed mainly of plants such as lavender and box, but later ones often included floral subjects. Some still exist.

Lag phase: Used to describe the first of five growth phases of most batch propagated cell suspension cultures in which inoculated cells in fresh medium adapt to the new environment and prepare to divide.

Lamellate: Having, consisting of, or covered with thin plates or layers.

Lamina: The blade of a leaf or petal, usually thin except in the case of the seaweeds where the fronds are often large and leathery.

Laminar air flow cabinet or hood: A structure in which a uniform flow of filtered air prevents settling of particles in the work area. The air is filtered through prefilter [furnace-filter quality], then a high efficiency particulate air (HEPA) filter that strains; out particles greater than 0.3 m. These hoods are commonly employed for aseptic plant tissue culture manipulations and are designed to accommodate one to the several persons, depending on the model.

Larva: The stage in the development of an insect between leaving the egg and transformation into pupa; commonly called caterpillar, grub or maggot. Usually it is the most destructive period of the insect life cycle.

Lath house: A structure, usually in hot climates, covered with narrow wood slats or brushwood to reduce the power of the sun while affording some protection against wind and slight frost. Lath houses are very popular in the southern states of America, in Mexico, and many other sunny areas of the world.

Lawn sand: Each manufacturer has his own formula but basically they all consist of a mixture of 3 parts sulphate of ammonia and 1 part sulphate of iron to 20 parts silver sand (all by weight). Mercuric compounds are often added to control moss. Lawn sand is best applied in spring or late summer at the rate 4 oz per square yard (100 g per square metre).

Lawn: A more or less flat area of ground covered with grass kept short by regular mowing. Lawns are a feature of most gardens in Britain and USA but are not so widely favoured elsewhere in the world except perhaps in parts of Australia,

Lax: Loose, relaxed, widely spaced.

Layer: A stratum of vegetation, as the shrubs in a wood.

Leaching: The loss of soluble (and some insoluble) minerals from the growing medium during the normal watering process. m rain. Some minerals leach faster than others; sometimes, producing an imbalance in the soil's chemical content with an adverse effect upon the plants growing therein. In general, leaching is more of a problem with pot plants than with those grown in the garden.

Leaf counting: Procedure used to time Easter lily flowering.

Leaf curl: A fungal disease of peaches, nectarines and almond trees in which the leaves curl up and thicken, and change colour from green to yellow and dark red. All affected leaves should be removed, and burned. A fungicide should be applied in spring, just before the buds begin to swell, and again in autumn when the leaves start to fall.

Leafhopper: A member, like the frog hopper, of the suborder Homoptera, often very colourful, usually quite small, and capable of jumping and flying. Many are pests and can be found on a variety of plants including potatoes and nettles, shrubs and grasses; they may also cause considerable damage in greenhouses. Control is by the use of pesticides.

Leaf gall: An abnormal growth, usually on the underside of a leaf, caused by bacteria. Attacked pot plants should be burned together with the soil in which they have been grown; in open ground where infected plants have grown, similar crops should be avoided for several years.

Light compensation point: That light intensity at which respiration and photosynthesis of the entire plant are balanced.

Light flux: The light intensity times the duration of light.

Light: Electromagnetic radiation with wavelengths in the visible range, 400-700 nm.

Ligneous: Woody or wooden.

Lignification: Hardening of tissue from wall deposition of lignins.

Limb: 1. A main branch of a tree or shrub. 2. The free or expanded part of a petal, sepal or leaf.

Limbate: Having a distinct or different-coloured edge or border, as in some carnation flowers.

Lime sulphur: A favourite fungicide and acaricide with Victorian gardeners for the control of scab on apples and pears, and against red spider mites on fruit bushes, strawberries, etc. Now largely superseded by modem chemical products such as Benomyl and Captan.

Lime-hater: A calcifuge. any plant unable to thrive in alkaline soil, e.g. many ericas.

Lime: A chemical compound supplying calcium to the soil and reducing its acidity. The amount of lime present in the soil determines whether it is alkaline, neutral or acid.

Liquid culture: The culture of plant cells on liquid medium, in suspension or on supports. Cultures are either held steady [stationary culture] or are shaken [agitated culture or shake culture].

Liquid medium: Medium not solidified with a gelling agent. Liquid media are used for suspension cultures and for a wide range of research purposes. They are also useful for Stage I cultures in some micropropagation protocols. Usually a support structure or wick is used to hold the tissue above the nutrient medium.

Long tom: A flowerpot much deeper than usual relative to its width; seldom seen today but popular prior to the introduction of plastic pots. Most, if not all long toms, were clay pots.

Long-arm pruner: Pruning shears attached to a long handle and operated by a remotely controlled lever mechanism.

Long-day plant: A plant which flowers only when nights are short and there are maximum daylight hours. Day length is often controlled artificially) by the use of supplementary lighting in commercial nurseries.

Macrobiotic: Applied to seeds that remain alive in a dormant condition, sometimes for several years, until a suitable environment prevails for their germination.

Macronutrient: An element, such as sulphur, phosphorus, magnesium, calcium, potassium and nitrogen, essential in large quantities for plant growth. Macronutrients are needed in concentrations of 1000 mg $g^{-1}$of dry matter or more.

Maggot: Larva or grub of many flies, especially the blowfly; frequently used to refer to the caterpillar of the Codling moth in apples. Often called a worm in USA.

Magnesium: An essential element in the production of chlorophyll in plants; a deficiency can inhibit normal development. Epsom Salts contain about 10 per cent magnesium and can be applied as a foliar feed or drench when dissolved in water and. diluted. An application once a year is usually ample to meet the requirements of most subjects.

Magnetic stirrer: An apparatus used for stirring solutions. A magnetic stir bar is placed at the bottom of the container and is twirled in the solution that requires mixing by an electrically driven, rotating magnet below the platform of the apparatus.

Maintenance pruning: The gradual removal of diseased and crowded branches from elderly fruit trees to encourage new and more fruitful growth.

Major element: An element which is essential for life and is required for the formation of the major structural parts of living organisms. They include carbon, nitrogen, oxygen, hydrogen, sulphur, phosphorus, calcium and, in some instances, silicon.

Male flower: A flower having functional stamens but no carpels.

Maleic hydrazide, MH: 1, 2-dihydro-3, 6-pyridazinedione: A synthetic plant growth retardant which is used as a herbicide against grasses. It causes a unique depression of growth, inhibition being marked but temporary. It inhibits germination of seeds, suppresses growth of roots and terminal shoots and retard flower and bud development. It prevents the formation of suckers in tobacco and tomatoes, and prevents the sprouting of onions and potatoes.

Malic acid: An acid in many fruits such as apples, grapes, etc.

Malodorous: Evil-smelling, stinking, as the foliage of Herb Robert when crushed or handled.

Manure: Formerly farmyard or stable refuse, now virtually any bulky material which supplies humus and plant nutrients to the soil, e.g. hop manure.

Marginal water plant: One that grows partially submerged in shallow water or in damp soil at the edge of a lake or pond.

Marketing phase: The movement of the plants and/or flowers from the forcing facilities to the wholesaler and/or retailer at the proper stage of development so that the consumer receives the maximum possible enjoyment.

Mast: The fruit of beech, chestnut and several other forest trees, mainly used for, or incorporated into, pig food.

Maternal inheritance: Inheritance controlled by extrachromosomal [cytoplasmic] hereditarydeterminants.

Mature: Fully developed or ripe.

MCPA: 2-Methyl-4-chloro-phenoxyacetic acid; a synthetic auxin used as a weed killer.

Mealy bug: Sap-sucking insect rather like a small woodlouse in appearance but coated with a white or pale grey waxy substance resembling meal. A greenhouse pest controlled by insecticides, especially systemics.

Medium formulation: One of many available, usually derived formulas that are used in plant tissue culture. The formulas commonly contain the macroelements and microelements [high and low salt formulas are available], some vitamins [B vitamins, inositol], hormones [auxin, cytokinin and sometimes gibberellin], a carbohydrate source [usually sucrose or glucose] and often other substances such as amino acids [the most common addition is glycine] or complex growth factors. Media may be liquid or solidified with agar the pH is adjusted [ca. 5-6] and the solution is sterilized [usually by filtration or autoclaving]. Some formulations are very specific in the kind of explants or plant species that can be maintained, some are very general.

Mercuric chloride or mercury bichloride [$HgCl_2$ mw 271:52]: This compound was once commonly used as an antiseptic and fixing agent [0.05-0.10%] for plant material. However, it is highly toxic and is now little used for this purpose.

Meristem culture: The culture of meristems; meristematic dome tissue without adjacent leaf primordia or stem tissue. In practice this usually means meristem tip culture. This may more generally imply the culture of meristemoidal regions of plants or of meristematic growth [associated with or sharing the characteristics of the meristem] in culture.

Meristem tip culture: Cultures derived from meristem tip explants, they are excised for virus elimination or axillary shoot., proliferation purposes, less commonly for callus production.

Meristem tip: A common explant comprised of the meristem [meristematic dome] and usually one pair of leaf primordia. Also refers to apical [apical meristem tip] or lateral [lateral or axillary meristem tip] origin. The term meristem tip is often confused with the term shoot tip, which is much larger and usually has more immature leaves and some stem tissue.

Meristem: A localized region of continuing mitotic cell division meristematic cells), of protoplasmic synthesis and tissue initiation. From these undifferentiated tissue new cells arise that differentiate into specialized tissues. Meristems are located at the apical [shoot and root tips] axillary, marginal or lateral [cambia] and other growing points. In addition, in vitro meristems may occur within more or less differentiated callus tissue and are termed meristemoids. The meristematic dome without any leaf primordial tissue is sometimes used as an explant in virus elimination work but usually meristem tip [meristcmatic dome plus one pair of leaf primordia] is the explant.

Microclimate or microhabitat: The climate in the immediate vicinity or surrounding an organism or its parts.

Microclimate: The climate of a small or limited surroundings or habitat e.g., the surface of the soil, or under the canopy of a fores: bioclimate.

Microcutting: A tiny cutting; as an explant such as a meristem tip removed for culture with the use of a dissecting microscope.

Microelement: A trace element; an inorganic substance that is essential to a plant in extremely minute amounts, such as copper, molybdenum, and zinc.

Micrograft: An in vitro graft. This procedure involves placing a meristem tip or shoot tip explant onto a decapitated rootstock, that has been grown aseptically from seed or micropropaged. The purpose may vary, from virus elimination, using meristem tips and virus-free rootstocks, to viral assays or to expedite grafting that is usually done in the greenhouse. The explanted scion is placed directly on the cambium of the severed stock or inserted into a T or inverted T-shaped epidermal incision on the stock. This technique has been used to eliminate viruses, viroids and mycoplasmas from Citrus species.

Microspecies: Generally only of interest to the specialist; a group of plants so closely resembling each other that they are classified under one species heading. The form *Rubus fruticosus* is often seen, meaning the group including all the microspecies.

Minimum inoculum size: The smallest inoculums that can be successful1y used for subculture.

Minimus: Extremely small variety of plant; the smallest of a group.

Mist propagation: A technique developed for the propagation of cuttings, especially hardwood cuttings, reducing transpiration and increasing the relative humidity by keeping them moist. A mist is produced by a series of fine jets placed at intervals of approximately 3 ft., switched on and off either by time control apparatus or an electronic leaf. The system is widely used commercially and small units are available for amateur use.

Mite: Any of several kinds of arachnid distinguished from insects especially by having eight legs and no antennae, e.g. red spider mite. Most are destructive pests controlled by chemicals called acaricides.

Miticide: Pesticide that is used to control mites.

Mitosis: The elaborate process of cell-division in plants.

Mitotic index [MI]: The ratio of nuclei undergoing mitosis [including prophase] to total nuclei. MI = [Number of nuclei in mitosis/

Total number of nuclei examined in the sample] X 100. A MI of 0.3 means that 30% of cells in the population are observed in mitosis.

Monocarpous: An ovary of a single carpel having a single ovary; producing only fruit.

Monoculture: 1. A culture of microorganisms, plant or animal cells that contain cells of only one species, strain or clone. 2. A method of farming of forestry in which large areas of land are planted with a single species of crop plant.

Monohybrid cross: A cross between two individuals heterozygous for only one pair of alleles; *cf* dihybrid cross.

Monohybrid inheritance: The inheritance of one pair of genes.

Monohybrid: An individual heterozygous for a single genetic factor.

Monomorphic: Having or exhibiting only a single form.

Mononucleotide: A nucleic acid basic-building-block. Composed of a pentose, a phosphoric acid plus a purine or pyrimidine base.

Moor: Area of unenclosed, often heather-covered, sparsely wooded land.

Mor: Humus formed under acid conditions; the opposite to mull.

Moraine: The gravelly detritus and debris on the edges and at the foot of a glacier, often where the glacier has retreated (e.g. after the Ice Ages) leaving the moraine in situ, usually with fresh water flowing under and through it at certain times of the year. In many respects similar to the scree found at the foot of mountains and cliffs but normally much drier. In nature plants grown in scree or moraine enjoy very good drainage and minimum humus. Moraine beds can be created in gardens with a supply of running water, otherwise a scree bed will be more suitable.

Morphogenesis: The origin and development of an organ or organism.

Mosaic: The symptoms of one of several types of virus to be found in horticulture. It can be spread by insects, birds and humans, and once infected there is no cure for a plant. Some viruses are benign and the infected plants are grown for their unusually attractive foliage, e.g. *Abutilon megapotamicum'variegatum'*.

Moss pole: A wooden or plastic pole to which sphagnum moss has been wired, used for training up a climbing or trailing plant; or a cylinder of wire netting tightly packed with sphagnum moss for a similar purpose. Sometimes called a 'totem pole' in USA.

Moth: The harmless mature stage of certain caterpillars; unfortunately the grubs that hatch out from their eggs are very destructive and may cause serious damage.

Mother bulb: That portion of the bulb that is currently flowering and producing a daughter bulb in the axil of a scale subtending the mother axis. The old mother scales encompass the new daughter axis.

Muck: Farmyard manure.

Mucro: A short sharp paint to tip, as at the apex of some leaves.

Mucronate: A type of leaf terminated by a short stiff point or mucro.

Mucronulate: The diminutive of mucronate; terminating in a very small mucro.

Mud: Wet soft soil or earthy matter, not solid but not quite liquid.

Mull: Humus formed under alkaline conditions; the opposite to Mor.

Multi-dibber: A piece of board with numerous small projections, dowels or pegs, used for making several small holes in a seed tray or bed simultaneously. Can be used for seed-sowing, pricking-out seedlings, striking cuttings, etc.

Muriate: Rough or warty, with short sharp points or prickles on the epidermis.

Mushroom compost: The material in which mushrooms are grown, consisting mainly of peat, composted straw and nitrogenous fertilizer; animal manures are seldom nowadays included in the formulae. Spent mushroom compost is widely available and is a useful addition to garden soils since it increases the supply of organic matter and water retention, but it has virtually no nutritional properties.

Mutagen: A chemical or physical treatment or agent capable of producing genetic mutation. Causes insertion, deletion or alteration of a base or part of the nucleic acid chain [DNA]. Artificial mutations can be induced in cells using chemicals or ionizing radiation. The process is mutagenesis. Electromagnetic and particle radiation, such as u.v., x-rays and b particle all increase mutation frequency. So do chemical mutagens such as ethylmethanesulphonate, acridine, nitrous oxide and many others

Mycotrophy: The condition of a plant living in symbiosis with a fungus.

2-naphthaleneacetamide acetic acid or 2-naphthoxyacetic acid or b-naphthoxyacetic acid (NOA or b NOA, $C_{12}H_{10}O_3$ mw 202.20). A synthetic hormone analog of the auxin type. It is sometimes used in plant tissue culture media and in horticulture to promote rooting of cuttings. It dissolves in base [ca. 1M HOH 1- or NaOH].

Nana: Small; dwarf. Usually applied to the diminutive form of a plant.

Naphthalene: Chemical obtained from coal-tar, etc. One of those used in the past for soil Sterilization but now almost superseded by pesticides.

Naphthaleneacetic acid (NAA): A synthetic plant growth substance with auxin-like activity.

National growmore: A general fertilizer produced to more or less standard formulae by several manufacturers. It first appeared during World War Two under the direction of the then Ministry of Agriculture and Fisheries but is still widely used by gardeners. It provides an equal balance of nitrogen, phosphate and potash in various concentrations.

Native: An indigenous species.

Natural break: A side shoot produced naturally on a plant, without pinching, pruning or any chemical treatment designed to induce breaking.

Natural cooling: Technique in which nonprecooled commercial bulbs are planted immediately on arrival and grown under cool natural conditions, but with frost protection, prior to being placed in the greenhouse.

Neb: Part of the handle of a scythe or the spout of a watering can.

Neck rot: Decay appearing at the neck of a bulb or corm, caused usually by a fungus which attacks a variety of subjects including onions, garlic and shallots.

Neck: The upper part of a bulb or corm where the top growth joins the base.

Necrosis: Symptom of plant injury, caused by spray damage, insect or disease injury, or other causes, which is characterized by dead, discolored cells and tissue.

Nematode: Small insegmented worm of slender cylindrical shape, resembling a piece of fine white cotton thread. Usually found in badly drained or poorly cultivated soils. An eel-worm is a typical nematode.

Neoplasm: Localized cell multiplication or tumour; a collection of cells which have undergone genetic transformation. These cells differ in structure and function from the original cell type.

Net veining, Net virus: A virus, similar to lace netting which produces symptoms in the leaves of some plants, e.g. *Pelargonium paltatum*. 'White Mesh'. In many cases the virus is not harmful and the effect is considerable attractive.

Netting: Garden netting can be made of natural materials, wire, or a variety of synthetics. It is used for many purposes such as support for climbing plants (peas, beans etc.), protecting crops from bird and animal damage, or as a wind-break. Galvanized wire netting is used mainly for making pens for animals or for barries such as open fencing.

Nitrogen cycle: The overall circulation of nitrogen through the biosphere. The cycle involves the nitrogen-fixing organisms (bacteria and cyanobacteria), organisms that produce ammonia from decaying matter, the nitrifying bacteria which oxidize ammonia to nitrite or nitrite and plants which assimilate nitrogen as ammonia or nitrate and form amino acids, as well as the denitrifying bacteria which convert nitrate back to nitrogen gas through nitrite.

No-digging: A method of cultivation which involves the application of large quantities of organic matter to the surface of the soil. No digging is involved at any time, and some remarkable results have been recorded.

No-soil composts: Peat-based products with a variety of chemical additives which can be adjusted to suit the plant being grown. First developed at the University of California, Los Angeles in the early 1940s to provide a uniform, dependable and largely disease-free alternative to natural soils No-soil or soilless composts are now widely used commercially in Britain and overseas. Aeration is particularly good but such composts are very light, lacking support for the plant and if allowed to dry out are very difficult to bring back to optimum water content. They all have a short life and plants in any peat-based compost will need regular feeding after a month or rapid deterioration can result. On the plus side, they are easy to handle, being much lighter than soil, and most plants respond exceptionally well to them until they reach maturity. The rapid depletion of the world's peat stock has resulted in the Increased use of coconut fibre as a substitute.

Node: The place, often swollen, where a leaf-stalk joins a stem, or where buds and side-shoots emerge from the stem.

Nodose: Having many nodes situated closely together; knotty.

Nodular: Having nodules or small nodes; of or like a nodule; in the form of nodules or knots.

Nodule: Small knot-like swelling on the root of members of the Leguminoseae family containing large numbers of a species of bacteria which attracts nitrogen from the air circulating around the roots, making it available to the plant and incidentally to other types of crop grown in the same ground in succeeding seasons.

Nog: The stump or snag remaining on a tree after a branch has been removed.

Nomenclature: The systematic classification of plants.

Non-vascular: Plants that lack vascular tissue, such as fungi, mosses, lichens, algae, etc.

Non-viable: Incapable of normal development, or survival.

Noncooled bulb: Bulb that is delivered direct to the forcer and has not received a cold treatment.

Notch graft: A plant graft in which the scion is inserted in a narrow slit in the stock.

Notch: An indentation or similar V-shaped incision.

Novirame: The fruiting or flowering shoot developing from a primocane in its second year, as in brambles (*Rubus*).

Nuclear stock: Plants produced vegetatively from virus-free stock, often from a single mother plant.

Nucieo-cytoplasmic ratio: The ratio of cell nuclear to cytoplasmic volume. This ratio is elevated in meristematic cells and low in differentiated cells.

Nursery graft: A plant graft in which the scion remains united with the stock only until roots develop on the scion.

Nurse: A culture technique or the callus upon which a filter paper is placed separating single cells from the callus in the paper raft technique. The callus release growth factors and nutrients that induce growth in a single cells supported by the filter paper (those being nursed) and sharing the communal environment.

Nursery bed: A piece of ground set aside for raising young plants until they are mature enough to be transferred to their permanent quarters.

Nursery stock: Hardwood ornamental and tree species being grown in a nursery for planting elsewhere.

Nutrient gradient: A diffusion gradient of nutrients and gases is set up in tissues where only a portion of the tissue is in contact with the medium. This is common in callus culture and causes differential growth rates and senescent regions of callus. Similarly diffusion gradients of nutrients are set up in the medium adjacent to the tissue. Gradients are less likely to form in liquid media.

Nutrient medium, Growth medium: A medium, liquid or solid, for the cultivation of microorganisms, cells, tissues or organs. Solid media are prepared from liquid media by the addition of gelling agent, e.g., 'gelatin (nowadays used only in special cases), silicic acid (when exclusion of organic compounds is necessary, or Agar-agar (used widely in bacteriology, usually at a concentration of 1.5 = 3%).

Nyctigamous: Used of flowers that open at night and close during the day; nyctigamy.

Nyctinasty: Response by plants to periodic alternation of day and night, e.g. opening and closing of many flowers, 'sleep-movements' of leaves. Related to changes in temperature (Thennonasty) and light intensity (photonasty).

Nymph: The Immature stage of certain insects including aphids, leafhoppers, etc. Nymphs normally resemble adults but are wingless and often have different colouring. A nymph is the distinguished from a pupa, which has its legs immobilized by being active.

Obcompressed: Flattened from front to back.

Obconical: Of a leaf, roughly conical but with the stalk at the narrow end.

Obcordate: Of a leaf, inversely heart-shaped; the reverse of cordate. being mutilated.

Obcuneate: Of a leaf, inversely cuneate and wedge-shaped, attached at the broad end.

Obdeltoid: Of a leaf, inversely deltoid, narrowest at the stalk end.

Obdiplostemonous: Having two whorls of stamens, the outer whorl opposite the petals, the inner whorl opposite the sepals.

Obelisk: An erect, usually square-sectioned masonry pillar, occasionally seen in landscape gardens and other public places, often with plants such as ivy growing up it.

Oblanceolate: Inversely lanceolate with the broadest part of the leaf above the centre, tapering toward the base.

Obligate parasite: A parasite entirely dependent upon its host for nutrition and therefore survival. Some host plants require a particular parasite for their well-being.

Oblique: Having unequal sides, as the leaves of Begonia rex.

Oblong-orbiculate: Of a leaf, basically oblong, almost as long as it is wide and appearing nearly round.

Oblong-ovate: Of a leaf, oblong with sides that curve in at either end.

Oblong: Of a leaf with almost parallel sides, and between two and four time longer than broad.

Obovoid: Of leaves and petals, roughly the shape of a hen's eggs with the stem at the narrow end; several fruit are obovoid, including some rose hips.

Obpyriform: Roughly pear-shaped with the stem at the narrow end.

Obsolescent: Almost obsolete, in the process of disappearing. Usually applied to an organ still evident but not functional.

Obsolete: No longer functional or fully developed; gone out of use.

Obtruncated: With the head removed; decapitated; being mutilated.

Obtuse: Of a leaf or petal that is blunt or rounded at the tip; obtusifolious.

Obverse: Of a leaf, narrower at the base or point of attachment than at the apex.

Occluded: A wound on a tree that is completely covered by callus; can also apply to other openings that have become closed off.

Occlusion: The process by which tree wounds callus over.

Ocrea Ochrea: A sheath formed of two stipules united around a stem.

Odd-pinnate: Of a leaf, pinnately compound with the pairs of leaves surmounted by a single terminal leaflet; sometimes called imparipinnate.

Odoriferous: Odorous. Sweetly scented; fragrant; emitting a pleasant smell.

Off-shoot: Methylester of a fatty acid that is used to kill shoot apices and promote lateral branching. Primarily used on azaleas.

Olitory: Old word for a kitchen garden or a kitchen gardener; from the Latin olitor- kitchen gardener.

Onion hoe: A small hand-tool with arched handle designed for weeding large onion beds in the days when this operation was performed mainly by women known as crawlers, from the fact that they worked on hands and knees.

Ontogeny: The developmental **[ontogenetic]** life history of an organism, its development from fertilized egg [zygote] to seed forming adult [reproductive phase], with emphasis on embryonic development.

Open continuous culture: A cell suspension culture with a continuous influx of fresh medium_maintained at constant volume by the efflux of cells and spent medium.

Open-centre: A method of pruning trees, particularly fruit trees, which leaves the centre clear of growth, and the remaining outer branches in the shape of a cup or vase.

Organ culture: The growth in aseptic culture of plant organs such as roots or shoots, beginning with organ primordi or segments and maintaining the characteristics of tile organ.

Organic gardening: Cultivation without the aid of chemicals of artificial origin.

Organic: Substances derived from the decay of leaves and other vegetation, animal waste and animal tissue.

Organized growth: The in vitro development of organized explants such as meristem tips or shoot tips, floral buds or organ primordia, or their denovo formation from **unorganized** tissues.

Organogenesis: The initiation [de novo] and growth of organs [roots and shoots, usually] from cells or tissues; as in organ culture. Organs may form on the surface of ex plants [direct organogenesis] or upon an intervening callus phase [indirect organogenesis].

Organoid: An anomalous organ-like structure formed in culture; as on leaves, roots or callus.

Ornamental garden: A loose general term for a garden designed for pleasure rather than utility, usually having formal flowerbeds separated by paths or grass verges.

Ortet: The original mother plam or donor plant from which vegetatively propagated plants are derived.

Orthotropism: The ability for the growth response of a plant to be orientated in line with a particular stimulus, such as the vertical growth of main stems under the influence of gravity.

Orthotropous: Ovules carried on a straight funicle (stalk), not curved or bent over.

Osier bed: An area of land where osiers grow and are harvested.

Osier: Any type of willow whose twigs are used for basket making, furniture, etc.

Osmium: tetroxide or osmic acid [O_sO_4 mw 254.20]. A fixing agent used to prepare tissues for electron microscopy.

Osmoticum: An agent such as glucose or sucrose employed to maintain the nutrient medium osmotic potential equivalent to that of the cultured cells [isotonic]. This prevents cell damage in vitro.

Over-potting: The transferring of plants from small pots into others too large for the size of plant, e.g. a plant rooted in a 3 inch pot being potted-on into a 10 inch container without any intermediate stages. In such cases the soil nearest the wall of the pot will have most of its nutrients leached out before die roots have extended into it; and in those cases where the chemical content is badly out of balance-bearing in mind that the basic chemicals (NPK) leach at different rates- the roots may never penetrate into the hostile environment and the plant will not achieve its optimum development.

Ovule culture: A culture derived from an explanted ovule; which may be fertilized in culture [in vitro fertilization]. This technique is used to study development of zygotes and young embryos and is sometimes used to rescue embryos susceptible to abortion when embryo culture is not possible.

Oyster shell: Crushed oyster shell was once considered to be an essential ingredient of bulb fibre. It contains a high proportion of calcium which is released slowly over a long period and, being porous, is able to absorb water and gradually release it as required. It has been almost completely superseded by sharp sand, perlite, etc.

Paper raft technique: A technique developed by W.H. Muir (1953) to promote development of single cells taken from suspension cultures in which cells are placed onto filter paper squares set on actively growing callus [nurse tissue]. Growth factors and nutrients from the callus tissue diffuse through the filter paper, promoting cell growth and development.

Parasite: Living organism (virus, bacterium, fungus, insect of plant) that draws sustenance from another organism; many house plants are susceptible to attack either by animal or plant parasites.

Parcel: Plot or area of land with recognizable natural or artificial boundaries.

Paring box: A simple device used to trim turves to a standard thickness. It comprises a three-sided frame mounted on a flat base, with wooden or metal vertical sides of the height of the desired thickness. Turves are placed in it grass side down and surplus soil skimmed off with a cutting blade or double-handled knife.

Parthenocarpy: The condition of producing fruit without seeds.

Parts per million (ppm): Equivalent to milligrams per liter.

Peat moss (moss peat): Partially decayed plant material often used as an ingredient in a growing medium: Usually acidic.

Peat-lite: Medium composed of peat moss and other ingredients, such as vermiculite and perlite.

Pedicel: Flower stalk.

Peruvian Guano: A valuable fertilizer found on the coasts of Peru, rich in phosphates and ammonia, the excrement of sea birds.

Pest: Any living creature which damages plants, except for fungi, bacteria and viruses, which are diseases.

Pesticide: A chemical substance for destroying or controlling pests, especially insects.

Pesticide: Chemical used to control undesirable organism. Growth regulators also are classified as pesticides, subject to EPA regulations and labeling.

Pestology: The study of harmful insects and other pests, and methods of control.

pH drift: A shift in pH that occurs as cultures grow and is related to the buffering capacity of the medium.

pH: The hydrogen ion concentration in the soil or water. The pH scale, which goes from 0 to 14, is used to measure acidity or alkalinity.

Photoperiod: The length of the day used in reference to its effect: on growth and flowering.

Photoperiodic response: Behavior of an organism to the length of day.

Photoperiodism: The response of an organism to the length of a light period. The most notable photoperiodic responses occur in plants where flowering, bud dormancy, leaf fall, tuber formation, etc. are controlled by day length. Plants are distinguished as long-day, short-day or day-neutral, depending on their response to variations in the period of darkness to which they are subjected. Responses in plants are the result of phytochrome-mediated phenomena; a reaction to red light may be reversed by a flash of far red light, and vice versa. Many animals also show photoperiodic responses.

Photorespiration: Most photosynthetic plants undergo this light-dependent respiration, differing from dark respiration. Its function is unclear. It involves the oxidation of glycollate to glyxylate in the peroxisomes, followed by recycling to a Calvin cycle carbohydrate: This process wastes CO_2, ATP and NADPH2 reducing photosynthetic efficiency.

Photosynthesis: The enzymatic conversion of light energy into chemical energy by forming carbohydrates and oxygen from CO_2 and H_2O in green plant cells. It is one of the most fundamental biological functions. In this process, the chlorophyll traps the energy of sunlight emitted as photons (i.e.. excitons) as transform it into chemical energy. This energy is stored in the chemical bonds that are produced during the synthesis of various food stuffs, the overall equation of photosynthesis is. Photosynthesis consists of two interconnected sets of reactions (i) Phase I reactions, and (ii) Phase II reactions. Phase I reactions take place in the membranes inside the chloroplast and phase II reactions take place in the open space (stroma) inside the chloroplast.

Photosynthetically active radiation (PAR): The flux expressed as moles or microeinsteins per meter squared per second (molm-2 or Em2S-1) or as watts per meter squared (Wm-2) over the wavelength range of 400-700 nm; the part of the light spectrum which is primarily absorbed by plants and used in photosynthesis.

Phytohormone: Plant hormone, a substance which is produced in cells in one part of the plant and is transported to other parts where it produces an effect.

Phytophthora: Genus of fungi that causes plant diseases, such as root and crown rots. A water mold.

Phytotoxic: Injurious or lethal to at least some plants.

Pick: 1. To gather fruit, flowers etc. 2. Abbreviated form of pickaxe, sometimes called pack.

Piece-root grafting: Grafting in which each piece of a cut seedling root is used as a stock.

Pig manure: Not so rich as other animal manures but provides humus and is a valuable fertilizer if left to rot for a year before use. If applied fresh, the ground should not be planted for at least six months.

Pinch: The removal of the shoot apex to overcome apical dominance and promote lateral shoot development.

Pinching: Removal or suppression of apical buds in order to stimulate to production of lateral branches.

Plant food: An organic substance (carbohydrate, fat, or protein) furnishing energy or building material for protoplasm.

Plant house: Usually a very large greenhouse.

Plant lice: A general term covering a variety of sap-sucking insects including aphids, blackfly, etc.

Plant tissue culture: Term applied to methods thatallow growth and development *in vitro* of plant cells, tissues, or organs, in or on nutrient media, usually under aseptic conditions.

Plant vase: A nontechnical term that describes the rosette of leaves that ascend forming a cup at the base of the plant. It is common to many genera within the Bromeliaceae family.

Plate budding: Plant budding by inserting a rectangular scion with a bud under a flap of bark on the stock in such a manner that the exposed wood on the stock is covered.

Plating efficiency: An estimate of the percentage of viable cell colonies developing on an agar plate relative to the total number of cells spread ontothe plate. Plating efficiency is a function of the tissue [species, explant and cell line]; medium composition; plating density and the phase of the stock culture.

Pleaching: The horizontal training of branches and twigs of trees, etc. to form a screen or windbreak.

Pleated: Folded lengthwise several times, as in some buds.

Pleiotropy: The condition in which several characteristics are effected by a single gene.

Plumule: The rudimentary stem of an embryo plant in a seed, and also in the germinating seedling.

Plunge: To sink a Pot plant up to the rim in soil, cinders, sand or similar materials. This method is often used to prevent the roots from drying out or to keep them cool in hot weather when the subject could be damaged by excessive heat round the roots. In commercial practice, bulbs required for Christmas flowering are plunged earlier in the year to ensure adequate root-formation before forcing can commence.

Polycarpic: Fruiting many times, or year after year.

Polyembryony: Condition where ovules have more than one embryo.

Post-emergence: Applied after a crop has emerged (also used in respect of weeds when so qualified).

Potassium, Potash: The K symbol of the NPK formulae, mainly responsible in plants for flower and fruit development. Originally called kanit or kanite. The word potash derives from the early crude method of leaching wood-ash and evaporating it in clay pots- hence pot-ashes.

Potato extract: A common [undefined] organic addendum to plant tissue culture media in monocot and another culture systems.

Potato-planter: An ingenious gadget for planting tubers and bulbs without stooping. It consists of a pair of tongs with flattened blades which can be separated after inserting in the soil, creating a hole into which the tuber can be dropped. As the implement is withdrawn, the soil falls in, covering the tuber.

Potting soil: Before the days of the John Innes Composts and soilless mixtures, this was the sieved soil sold by nurserymen for potting, usually dug from a comer of the nursery and often unsterilized.

Potting bench: A table reserved for potting-up plants, usually in a potting shed or greenhouse. Most potting benches have three raised sides to prevent the potting medium spilling, and quite often a shelf above and below to house pots, labels, tools, etc.

Potting shed: A building used for all potting operations and for storing garden tools, seeds, etc.

Prefilter: A coarse filter [furnace filter] such as those used in a laminar air flow cabinet to screen out large particles before air is forced through a much finer filter [HEPA filter].

Pre-planting (sowing): Applied before planting (sowing) a crop.

Prick out: Transplant seedling or cutting to larger container or into pot.

Prop root: A rigid aerial root that emanates from the stem, descends to the ground & helps to support the main stem, as in some trees.

Propagating frame: Normally an unheated outside structure of glass and wood used for raising young plants that need slight protection.

Propagating house: A greenhouse or part of a greenhouse set aside entirely for the raising of young plants, whether from seed or cutting; often specially equipped with heated benches and misting sprays.

Propagation: Methods by which plants are increased in number, such as grafting, division, cuttings, seeding, air-layering, budding, etc.

Propagule: A plantlet or offset produced naturally by an adult plant without assistance or special treatment. Such as *Tolmeia menziesii* which produces buds that develop and form new plants at the base of the leaf blades.

Protoplast culture: The isolation and culture of plant protoplasts by mechanical means or by enzymatic digestion of plant tissues or organs, or cultures derived from these. Protoplasts are utilized for selection or hybridization at the cellular level and for a variety of other purposes.

Protoplast fusion: The coalescence of the plasmalemma and cytoplasm of two or more protoplasts in contact with one another. Initial adhesion is a random process but coalescence may be promoted in various ways [induced fusion]. When adhesion occurs between adjacent protoplasts during enzymatic wall degradation or between freshly isolated protoplasts in the absence of a fusion agent, it is termed spontaneous fusion.

Prunning: Seasonal or periodic cutting back or removal of branches and stems, especially when dead, diseased or overcrowded, to improve health and general appearance, restrict excessive growth and promote flowering.

Pseudobulb: A swollen growth of bulb-like appearance, but not a true bulb. common in some orchids.

Pseudobulb: Swollen stem of an epiphytic orchid often developing at intervals from an underground rhizome and bearing leaves and flowers.

Puffer: A dusting device usually incorporating some form of bellows for applying powdered fungicides and pesticides to plants.

Pullulate: To sprout or develop by sprouting, as of a seed or bud.

Pulp: The fleshy part of fruit, often the edible part.

Pulse: The edible seed of leguminous plants such as peas, beans, etc.

Pulverulent: Literally, dusty as applied to the powdery coating on the stems and leaves of some plants. It occasionally occurs in specific names such as *Primula pulverulenta*

Pulvinate: Cushion-shaped, forming a dense low tuft or dome.

Punctum: A speck or dot, such as a spot of colour or a small depression on a surface.

Puncture: A small hole or perforation, especially in a leaf, caused by a biting insect or grub.

Pungent: 1. Acrid to taste or smell. 2. Sharp; ending in a hard sharp point or tip.

Punk: Rotten wood or the fungus that grows thereon; an expression mainly used in USA.

Pure culture: Axenic culture.

Pure line: All cell or individual members are homozygous for one or more characters or genes and will give rise to more cells or organisms with the character(s) under consideration.

Pustular, Pustulate: Blistered or apparently so; warty; pimply.

Quadrat Chart: A chart of squared paper on which the results of counting the species in a quadrat are marked, used mainly to record the investigation of large areas. A rectangle of land is marked out with tapes, then divided into smaller squares repeatedly until manageable-sized quadrats are produced and numbered to correspond to the squares on the quadrat-chart.

Quadrat: A square frame of metal or wood, or the area of vegetation enclosed by it, used to calculate the density of species growing on a certain area of land. Both the device and the area it encloses are referred to as a quadrat.

Quadrigeneric: A hybrid whose ancestry includes four different genera.

Quadripartite: Of a leaf, deeply cleft into four parts.

Quagmire: A boggy or marshy area of land; fen; slough; quaking bog.

Quantitative long-day plant: Plant that is not completely inhibited from flowering by short-day treatment but hastened by long-day treatment.

Quarantine: The control of plant import and export for purposes of disease and pest control, and to prevent spread of disease and pests. Control may involve isolation of plant material, sometimes for extended periods of time, to determine its disease status.

Quarender: Early type of red apple common in Somerset and Devon in Victorian times.

Quassia: An old-fashioned insect deterrent and bird repellent manufactured from the wood of the *Picrasma excelsa* tree. It has been largely superseded by modem chemical products but is still sometimes combined with derris as a safe insecticide, harmless to all warm-blooded animals, including humans.

Quatrefoil: A four-petalled flower or a leaf consisting of four leaflets.

Queach: A thicket or dense hedge,

Queening: A type of apple, seldom seen today:

Quercetum: A group of collection of oak trees; oak woodland or forest. A plantation of oak trees.

Quicklime: Oxide of lime, the product obtained by burning chalk or limestone in a kiln.

Quiescent centre: The central core of a root-apex which lies between the root-cap and the meristem.

Quiescent: Quiet, at rest, not necessarily dormant and having the potential for resumed activity; can apply to cells unlikely to divide, the non- meristematic cells.

Quilled: Of petals which are tubular for all or much of their length, such as certain dahlias often called cactus-dahlias.

Quince: A rosaceous tree (*Cydonia oblonga*) or the large, round, acid fruit it produces, which are commonly used in the preparation of jams, jellies, marmalades. In India it is grown in Kashmir on small scale.

Quincunx: An ornamental method of planting trees one to each corner of a square with a fifth in the centre; a favored system with a fifth trees in the past but now largely discontinued.

Quinsy-berry: Old country name for the blackcurrant, probably derived from its use in folk medicine to relieve the symptoms of throat infections.

Quitch-grass: Alternative name for couch-grass, a troublesome weed with creeping rootstocks, also called quack-grass.

Race: A vague term for a group of plants connected by a common ancestry; a subdivision of a species.

Radial: Of flowers with similar petals which radiate symmetrically from a common central point.

Radiate: Spreading from a common centre point.

Radiation: Rays of heat, light or particles in wave form.

Radical: 1. Basal leaves; those proceeding from the base of a stem or from a rhizome or crown. 2. Arising from or near the root.

Radicant: Rooting from the stem, e.g: ivy.

Radicate: 1. To plant firmly. 2. Deeply rooted.

Radicel: A rudimentary root or rootlet; that part of a seed which will become the primary root.

Radiculose: Having many rootlets.

Radius: 1. Outer edge of a composite flower head, e.g. daisy (*Bellis perennis*). 2. A radiating branch of an umbel.

Raffia: Fiber derived from the leaves of the raffia palm which forms a flat strip some 4 ft or so in length, used in the past for tying and training plants. Now largely superseded by string and plastic strip, it is still preferred by many gardeners for such purposes as grafting and budding.

Raft: A length of bark or similar material, usually oblong, on to which epiphytic plants can be wired. Widely used in the cultivation of epiphytic orchids.

Raised bed: A plant bed raised above normal ground level and contained inside a wall of peat, bricks, stones, etc. Raised beds are easier for disabled people to manage, and enable a large number of subjects to be grown in a small space.

Rake: A long-handled implement with a number of prongs or metal dowels set into a cross-head at right angles to the handle and used for breaking immune, up the soil surface into a fine tilth ready for sowing. Garden rakes can be made of metal, plastic or stiff rubber; hay rakes are usually much larger and made of wood metal cross-braces. Small handrakes for close work have a short handle and three or four tines bent sharply at right angles.

Rambler, Rambling: Straggling; wandering in a disorganized manner; Climbing and sprawling irregularly. A rambling rose is in nature an uncontrolled spreading shrub, whereas a climber tends to grow mainly upward in a more controlled manner.

Ramentum *pl*: Ramenta: Thin scaly membranous tissue on the surface of leaves and stems.

Ramet: The vegetatively propagated offspring of an ortet. An individual of a clone.

Ramification: The arrangement of a tree's branches, the pattern of their spreading.

Ramifying: Branching; dividing into branches.

Recessive: An inherited character not apparent in the organism which inherits it although it may appear in future generations. A recessive gene controlling such a character needs to be present in both members of a gene pair in order to express the character.

Reciprocating shaker: A plateform shaker with a back and forth action used for agitating culture flasks at variable speeds.

Recombinant DNA: Genetic material with novel gene sequences produced by cross-overs, chromosome reassortment, by other natural means or through genetic engineering.

Reconstructed cell or recon: A viable cell hybrid, cybrid or a tnmsformed cell resulting from genetic engineering.

Reproduction: The generation of new individuals by sexual, non-sexual or vegetative means.

Resistance transfer factor: A plasmid, present in certain types of bacteria such as E. coli, that can impart resistance to antibiotics in animals exposed to them.

Resistant: Having powers or properties of resistance. Some plants are selected or specially bred to be resistant to specific diseases, e.g. rust onchrysanthemums. This renders them better able to ward off an attack but does not mean they are immune, a term applied only to a plant which is totally protected, whether naturally or by breeding.

Response group: Classification of cultivar based on response to environment; for chrysanthemums, indicates number of weeks required to flower after start of short days. For snapdragons, indicates time of year cultivar should been flower.

Reticulate venation: Finely netted or net-like. In the case of the leaves of *Pelargonium peltatum* 'White Mesh' the virus emphasizes the reticulate venation which is already present. In grasses, where there is no main vein with branches, it is known as parallel venation.

Rhizobium: A type of bacterium found in the roots of leguminous plants connected with the fixation of nitrogen. One of the many beneficial bacteria.

Rhizome: Normally a horizontal creeping stem lying on or under the ground from which shoots will arise and roots descend; a type of rootstock. The bearded iris is an example of an underground rhizome.

Rhubarb pot: A fairly tall earthenware vessel, open at the bottom and the top and fitted with a lid. Victorian gardeners placed them over rhubarb crowns in early spring to produce long tender stems. Today, rhubarb is usually forced commercially in light-proof sheds, but pots can occasionally be seen in private gardens, and are much sought by collectors of antiques.

Rib: The primary vein in a leaf; can also be a ridge on a stem or fruit.

Ring culture: A method of growing certain types of plant e.g. tomatoes, by which they are encouraged to produce a dual root system, one for nutrients, the other for water. Very popular 30-40 years ago but not so well thought of today.

Ringing: A method of restricting the growth of trees, particularly fruit trees, to increase cropping. A ring of bark not more than 6 mm. wide is removed from the trunk about 60 cm above ground level in spring; this will callus over as the season progresses. A wider ring may well cause the death of the tree. In partial ringing, instead of a full ring being cut around the tree, two half-rings are removed on opposite sides of the trunk and about 15 cm. apart.

Ripe: Of fruit, fully developed, having mature seed and so usable as food.

Rond point: The centre point of a formal garden layout from which avenues radiate. It is usually a circular flowerbed or paved area.

Roof garden: A garden created on a flat roof, often above a block of flats or a departmental store, to which all materials and supplies have to be elevated.

Root aphid: Sap-sucking insect which attacks roots of some vegetables and flowers. Control is by watering the plants with a suitable insecticide.

Root ball: The matted roots of an established pot plant, tree, shrub or any other plant, including the soil enclosed by the roots.

Root crop: Plants grown for their roots; many vegetables, including potatoes, turnips, carrots, etc. Onions are not a root crop because the roots are not eaten,

Root culture: Isolated root tips of apical or lateral origin may produce in vitro root systems in determinate growth habits. These were among the first kinds of plant tissue cultures [White. 1934] and remain important research tools in the study of developmental phenomena; mycorrhizal; symbiotic and plant-parasitic relationships.

Root dipping: Immersing the roots of young plants in a solution of insecticide to protect against certain pests. Brassicas, for instance, are dipped or puddle as a protection against club root disease.

Root restriction: A method of encouraging more prolific flowering or fruiting, forcing the plant to restrict its growth and concentrate its energies elsewhere. Some subjects do better if their roots are restricted, including many bulbs and several kinds of pot plant.

Root rot: Condition often caused by excessive watering and water logged soil but may also be the result of attack by several kinds of disease organisms such as fungi.

Root run: The area of soil occupied by the roots of a plant.

Rooting hormone: Compound such as IBA or NAA that stimulates rooting.

Rooting room: A controlled temperature faculty used to root and satisfy the cold requirement of bulbs.

Rosary: A rose garden, or a rose bed in a mixed garden; a favourite expression in Victorian times but seldom heard today.

Rotating culture or rotator: A wheel-like device for slowly (ca. 1 rpm] rotating and gently agitating cultures, usually in a vertical plane.

Ruderal: Of a plant, growing in waste places or among rubbish near habitation, e.g. rosebay willowherb (*Epilobiun angustifolium*).

Rust: A plant disease showing as rust-coloured spots caused by fungus spores. A portmanteau word covering a number of different fungi; many of which include specific hosts, such as wheat rust, chrymnthemum rust, pelargonium rust, etc.

Rustic poles: Timber in its natural state, usually still carrying its bark, used for constructing pergolas, arches, fencing, garden furniture, etc.

S phase: The cell cycle phase during which DNA synthesis occurs.

Saddle-graft: A method of grafting used when the stock and scion are of the same circumference. The end of the scion is notched in the form of a 'V' and the stock shaped to fit it; they are then bound together and sealed. An inverted saddle-graft is when the stock has the 'V' cut in it and the scion is pointed to fit.

Saline: Salty; impregnated with salt as seashore sand, etc.

Sampler: A young tree or sapling left standing after adjacent trees have been cut down, especially in forestry.

Sand: An ingredient, to a greater or lesser extent, of most soils, which, although having no food value, is nonetheless useful for aeration. If used in potting mixtures it must be sharp sand; soft builder's sand tends to clog and consolidate the Soil.

Sap-sucker: Any insect that sucks the sap of a plant, such as an aphid.

Sapling: A young tree; a tree at any stage before the heartwood has hardened.

Scarifier: A machine or implement with prongs for loosening the surface of the soil; a particular type of rake used for removing moss from lawns.

Screen house: A metal or plastic mesh enclosure similar in function to a greenhouse but with more exposure to the elements.

Screenings: Material which has been passed through the screen or sieve, or the residues from screening.

Scurf: The scaly covering of leaves and stems of bromeliads which gives them a silvery appearance; technically known as lapidate leaves or stems.

Sealing agent: Grafting clay, grafting wax, etc. applied to freshly cut or damaged surfaces to prevent the entry of fungus spores, Wood-boring. insects, etc. A sealing agent must be. water-proof and elastic so that the plants natural callusing process can proceed normally underneath.

Second soil: The soil remaining after seedlings have been planted out, or from pots when the plant has died. This soil may contain disease and should not be reused; it should be put on to the compost heap.

Seed bed: An area of land specially prepared for seed sowing, usually by raking to a fine tilth.

Seed box: A rectangular container in which seeds can be raised. The standard British seed tray is about 35.5 x 21.5 X 5-7.5 cm. deep and they are obtainable in wood or plastic.

Seed dressing: Treating seeds before sowing to deter attack by soil-borne and other pests. In the past peas and beans were dipped in paraffin, then coated with red lead to discourage birds and mice, a practice which is now illegal.

Seed drill: I. A mechanical implement used for the efficient sowing of seeds. Especially when large areas, such as cornfields, arc involved. The seeds are contained in a hopper and released in a regulated manner as the wheels rotate. Smaller hand-held gadgets are available for gardeners but these tend to be more hit-and-miss. 2. A furrow drawn in, the soil to receive seeds.

Seed head: Any usually dry fruit that contains ripe seeds.

Seed leaf: Also known as Cotyledon. The first leaf or leaves-produced by a seed after germination. In some plants, e.g. peas, the seed leaves remain below ground.

Seed viability: A seed's capacity for living or germinating. Very few seeds have 100 per cent viability, and it lessens as the seed ages. The, seeds Act of 1920 requires all vegetable seeds to conform to prescribed minimum germination percentages, but flower seeds are not covered by the Act.

Seedling: 1. The young plant that develops from a geminating seed. 2. A plant raised from seed as distinct from a cutting or scion.

Self-Incompatible: Plants incapable of self-fertilization, often through the incomplete development of the pollen tube

Semi-ripe: Type of cutting used for the propagation of many shrubs and similar woody plants such as heaths and heathers.

Semicontlnuous culture: The maintenance of in a culture vessel in an actively dividing state by periodically draining the medium and adding fresh medium.

Senescence: The ageing process in (plants and animals); the changes that occur between maturity and death. Not all parts of an organism age at the same rate, and in deciduous trees the loss of senescent leaves in autumn is a normal process related throughout the life-span of the tree.

Shot-hole disease: A fungus disease, or the summer stage of bacterial canker, both of which affect *Prunus* species. Control is by the use of copper-based sprays at precise times of the year.

Shy flowering: Producing few flowers; backward; reluctant to flower.

Sickle: Hand-held reaping hook with a short handle and almost semi-circular blade.

Side graft: A method of grafting whereby the scion is inserted into the side of the stock, and after it has taken the stock is beheaded just above the union.

Side-dressing: A top dressing or fertilizer applied alongside growing crops without actually coming into contact with the foliage or stems of the plants.

Slag: The waste product from iron furnaces which can be used as a general fertilizer; usually marketed as basic slag.

Slug: Slimy land-moll use, very destructive, especially to young plants but also to some root crops. They differ from snails in having no visible shell although some have a reduced internal shell.

Slugworm: The larva of a species of sawfly which resembles slugs but has legs. They attack the foliage of plants and can be controlled by a variety of insecticides.

Smut: Fungus disease, mainly on cereal crops but also found on some garden. plants; the black powdery deposit caused by this.

Softwood cuttings: The method of propagation most used for perennials, sub-shrubs, alpines and certain conifers. The young shoots green shoots are taken trimmed and inserted into a propagating bed where they normally root rapidly. Hardwood cuttings are taken from older riper wood and will take longer to root.

Soil block: A usually rectangular or cylindrical block of compressed soil, used for sowing seeds individually. Blocks can also be made from peat or pear-based composts.

Soil conditioner: Any substance incorporated into the soil to improve, its physical structure; can be organic or chemical; can also include substances which improve drainage or aeration whilst providing no additional nutrients, such as sand and grit. Many soil conditioners marketed today are seaweed derivatives or gypsum-based.

Soil mark: The mark or stain on the stem of a plant indicating the soil level before it was moved or transplanted.

Soil: The upper strata of the ground in which plants grow.

Solar propagation: A method of rooting plant cuttings involving the use of a modified hotbed; bottom heat is provided by radiation of stored solar heat from bricks of stones in the bottom of the hotbed frame.

Source done: A source that originates from a single plant or explant within a clone; as from a specific virus tested [SVT] or specific pathogen tested [SPT] individual explants or plant.

Source plant: A mother plant or donor plant from which an explant used to initiate a culture is taken.

Special cooling of tulips: A special technique used to program tulips for cut flower forcing. When used, the entire cold requirement is given as a dry cold treatment and rooting takes place during the greenhouse phase of forcing. Also known as "41°F forcing" and "direct forcing".

Splice grafting: A simple method .of joining a scion and stock which are equal in girth.

Split application: The total amount of plant growth regulator divided between two or more applications made at different times to the same crop.

Split stone: A condition of stone fruits, especially plums, in which the exocarp splits to reveal the kernel or seed.

Spontaneous fusion: Uninduced protoplast fusion which may occur between freshly isolated protoplasts or following adhesion of adjacent cells during enzymatic cell wall degradation.

Spontaneous variation: The variation in plant populations derived from tissue cultures not exposed to mutagens but occurring as a result of the culture conditions.

Sport: A mutation, whether induced or accidental. A plant that varies spontaneously from the normal, either in leaf or flower form.

Sprayer: Any device used for applying insecticides, chemicals or plain water to plant foliage. The old type of syringe has now been superseded by smaller and held plastic models, or by the larger units incorporating a pump-action pressurized tank and lance

Sprinkler: Any automatic device for applying water to garden or lawn in the form of a fine spray.

Staminate: Male, having stamens but not pistils.

Stem rot: A general term applied to any disease which attacks the stems of plants. Sometimes called foot rot, damping off, black rot, etc.

Stem topple: In tulips, the physiological disorder that is characterized by the collapse of a small portion of the intermode of the floral stalk located just underneath the flower. It can occur either just before or during flowering.

Stock plant: A mother plant from which cuttings or scions will be taken for propagation purposes.

Stone fruit: A drupe such as cherry or plum.

Stone pit: A virus disease of pears resulting in fruits having small, gritty spots in the flesh.

Stubble: The short lower ends of grain-stalks left in the ground after harvest.

Stump: 1. The base of a tree left in the ground after felling. 2. The stem of a plant after the head has been removed, e.g., cabbage.

Stunt: To check the growth of a plant by chemical or physical means. e.g. bonsai

Subtropical: Plants originating in subtropical of the world, usually grown in heated greenhouses but some may be planted outside in favored parts of Britain including the south coast of Devon and the west of Scotland.

Suckering iron: A metal tool specially designed for removing suckers from the rootstocks of roses and other bushes.

Surfactant (Surface-active agent): A substance which reduces tension, thereby increasing emulsifying spreading and wetting properties of liquids.

Suspension culture: Cells and groups of cells [aggregates] dispersed in an aerated, usually agitated, liquid culture medium. These are obtained by adding friable callus to the medium. The plant species explants type and treatment; the composition of the nutrient medium; and many other features determine the size and nature of the cell aggregates, which may appear different during each culture growth phase. These cultures are used to study cell division, differentiation and metabolism, in secondary product synthesis or form the basis for single cell lines, callus cultures, somatic cell embryogenesis and many other culture purposes.

Sweal: To scorch or burn off; as heather, gorse etc.

Syringing: Application of water, sprayed on the foliage. Used to reduce transpiration, control some insect pests, reduce leaf temperature, or as a means of watering.

2,3,5-triiodobenzoate (TIBA mw 499:81): An inhibitor of auxin movement or transport [antiauxin], sometimes included in plant tissue culture media for its growth promoting effects.

2,4,5-T: 2,4,5-trichlorophenoxacetic acid, a translocated hormone weed-killer use to control scrub and woody vegetation.

Tamping: Firming; gently compacting.

Tang: 1. A prong, or tine, or a garding fork. 2. The tapering end of a file or trowel which is secured in a handle.

Tank mix: Two or more herbicides or other chemicals mixed in the spray tank prior to application.

Tar disc: Impregnated disc or tarred felt or thick paper about 3 inch in diameter with a radial slit and central hole. Widely used in earlier times for the control of cabbage root fly, now largely superseded by chemical insecticides.

Tar oil wash: Otherwise known as winter wash because it can be applied only during the dormant period of deciduous trees or the foliage will be severely scorched. Used to control a variety of pests such as aphids, scale insects, caterpillars, etc.

Tedder: An implement for turning over hay or grass, usually towed behind a tractor or horse.

Test tube fertilization: Pollination, followed by fertilization, occurring in vitro.

Thrip: A tiny sap-sucking insect of the order Thysanoptera with four hair-fringed wings. Many are injurious to plants. Otherwise known as thunder bugs or thunderflies because they are more troublesome in hot weather. Thrips is both singular and plural.

Tiplasmid: A portion of the genome of the bacterium Agro bacterium tumefaciens, the agent involved in crown gall disease. This plasmid is a Useful [experimental vector for the transfer of genetic information into plant cells.

Topping: Removing the growing tip of a plant to force the production of laterals, also removing the apex of broad bean plants to control black fly infestation.

Tortrix moth: Small moth from whose eggs leaf rolling caterpillars are produced.

Tortuous: Having many twists and turns; contorted, as the stems of the twisted willow *Salix tortuosa*.

Trace nutrients: a general term for any essential dietary component required in small quantities like Trace elements and vitamins deficiency leads to deficiency symptoms, e. g., I-vitamin deficiency diseases.

Tracer: A substance that can be followed within a reaction or an organism; as radioactive isotopes and certain dyes.

Transpiration: The continual evaporation of water, from leaf and stem surfaces. The rate of transpiration is affected by several factors such as temperature, light, humidity, etc. It is controlled to an extent by the guard cells of the stomata.

Transplant shock: Refers to the ,stress involved when Stage II, or Stage, III cultures are transplanted to soil [Stage IV]; many or all of the regenerated shoots or plantlets may die if suitable care is not taken to acclimatize them gradually to the soil environment, particularly prevention, of water stress.

Transplant: 1. To relocate or remove to a new growing place. 2. The cultured tissue or explants relocated or transferred to a new site [in vitro]. 3. Stage IV; the transfer of plantlets or shoots ex vitro from asceptic culture to soil. The process is transplantation.

Transplant: Moving a growing plant to a new location when sufficiently developed.

Trapeziform: Four-sided, with only pair or sides parallel.

Tree banding: The fixing of grease bands around the bole or main branches of a tree to trap crawling ascending insects. Sticky bands serve the same purpose as do bunches of hay or straw tied around the tree trunk, but these must be inspected frequently and the trapped pests destroyed.

Trembling bog: Swampland over water or soft mud shaking at every footstep; also known as a quaking bog.

Trench: 1. Long narrow slit cut out of the ground. 2. A method of digging or ploughing so as to bring the lower soil to the surface.

Trenching: A method of digging to a depth of nearly 30 inch by removing the soil in three spits whilst retaining the various strata at their original levels.

Trickle irrigation: A system of automatic watering and feeding whereby the fluids trickle or drip slowly over a long period into the compost in which the plant is growing.

Tropical: Of plants that originate in the tropics and need an optimum temperature of 70°-80° F.

Tubule: Minute tubular structure in a plant.

Tumor inducing principle (TIP): The plasmid carried by Agrobacterium tumefaciens, the crown gall organism. Through incorporation into the host genome the host tissue is transformed into tumor tissue.

Tump: A hillock or clump; a mound, usually grass covered

Tunic: The dry, papery scales that surround fleshy organs of a bulb or corm.

Tunicate: Having several concentric layers, as in onions.

Turion: 1. An underground bud on a rhizome from which a new stem will arise. 2. A young shoot or sucker. 3. An overwintering bud on certain water plants.

Turk's cap: A flower that resembles, an ancient Turk's cap, e.g. the Pyrenean lily (*Lilium pyrenaicum*) in which the petals are severely inflexed.

Twig: 1. A small shoot or branch; a side growth. 2. A water divining rod.

U Cordon, Double U Cordon: Methods of training fruit trees and bushes, more often used with soft such as red and white currants.

Ubiquitous: Occurs everywhere, as do bacteria in the environment.

Uliginous: Growing in muddy or swampy areas, a bog plant.

Ulmification: The process of peat formation.

Ulmin: A gummy substance which exudes from certain trees, particularly the elm (*Ulmus procera*).

Umbellate: Resembling an umbel or carried in an umbel; umbrella like.

Umbellet: A secondary umbel in a compound umbel.

Umbelliferae: A dicotyledonous family and commonly known as the carrot family. Its members are distinguished by their characteristic inflorence, the umbel, a feature that made them one of the first natural families to be clearly recognized.

Umbelliferous: 1. Any plant having an umbel or umbels. 2. A member of the family Umbelliferae.

Umbelliform: Resembling an umbe1.

Umbellule: A partial umbel.

Umbillicate: Having a depression, usually central; navel-like.

Umbo: A knob or projection, occasionally found on the caps of some fungi.

Umbonate: Having a central boss.

Unarmed: Lacking thorns or prickles.

Uncinate: Hooked at the end, or having a hook: hook-shaped.

Uncus: A hook or hook-like appendage.

Underground stem: Any of the stems, that grow underground and are often mistaken for roots; principal kinds are rhizomes, tubers, corms, bulbs, and rhizomorphic droppers.

Undergrowth: Growth of shrubs and plants under trees, often dense or matted.

Understock: The lower portion of a budded or grafted plant that develops into the root system.

Understorey: A ground cover of shrubs often found underneath the tree canopy in tropical rainforest and deciduous woodland.

Underwood: Small trees or shrubs growing beneath larger trees. Brushwood.

Undulate: Wavy; of leaves or petals with wavy surfaces or edges, usually applied to those that are wavy up and down; those that are wavy in and out are termed sinuate.

Unguis: The claw of a petal.

Unifoliate: Having a single leaflet but compound in structure.

Unilateral: One sided

Unilocular: Of ovaries with a single cell or chamber.

Union: The joint where a scion has been grafted on to the rootstock, usually indicated by a slight swelling.

Uniseriate: Arranged in a single row or layer.

Unorganized growth: In vitro formation of tissues with a few differentiated cell types and lacking recognizable structure; as with many calli.

Upright: Habit of a plant with vertical or semi-vertical main branches or stems.

Uptake: The process of taking water and nutrients from the soil into the roots of a plant, or a taking substances into a cell or organelle.

Urceolate: Pitcher-shaped.

Uredo: Rust in plants; rust-fungus in its summer stage.

Uredosori, Uredia: Clusters of uredospores.

Uredospore, Urediospore: The rust spore must familiar to gardeners, responsible for the rapid proliferation of rusts during the summer months, which is why they are often called summer spores.

Utricle: 1. A bladder or bladder like envelope of certain fruits, usually enclosing a single seed 2. A bladder on bladderwort, which contains air to give the plant buoyancy.

Uva: A grape like berry formed from a superior ovary.

Vapour drift: Movement of herbicide in the form of vapour from the treated area after application.

Variegated: Of a leaf marked in two or more colours.

Variety: Subdivision of species which, when produced by special horticultural techniques in known as a cultivar; the third word of the Latin or scientific name denotes the variety concerned.

Vase life: Longevity of cut flower.

Vermiculite: Mica platelets, formed by heating to about 1360oF and used as and ingredient in growing media.

Verticillium: Genus of fungi that causes vascular disorders of plants.

Volatility: Means the property of a substance or substances to convert into vapor or gas without chemical change.

Volume rate: The amount of liquid applied per unit area. Spray volume rates are often described as high, medium, low, very low or ultra-low (ULV), but to avoid confusion they are best specified in litres per hectare.

Vacillate: To sway to and fro; unsteady; in need of support.

Vacuolation: The formation of vacuoles during the development of living cells.

Vacuole: Space within protoplasm usually containing a fluid or air.

Vagina: A sheath; a sheathing leaf-base.

Vaginate: Sheathed; having a sheath

Vaginule: A little sheath, especially one surrounding the base of a moss seta.

Vaporizer: A heated container containing a vaporizing chemical which releases pesticides into the greenhouse atmosphere, usually electrical but not always so.

Varicoloured: Diversified in colour, same as variegated.

Vegetative propagation: The method of propagation using pieces of living plants as opposed to seed. In vegetative propagation the new plant will carry all the characteristics of the parent. All plants vegetatively produced from one parent are called a clone.

Vegetative reproduction: A form of asexual reproduction in which specialized multicellular organs formed by the parent become detached and generates new individuals.

Veil: The ring of tissue which in the early stages of development of a mushroom or similar fungus connected the stipe with the cap and which ruptures as the fungus matures.

Vernalization: 1. Low temperature exposure required by some plants to induce bud break or flowering. 2. Low temperature exposure of seeds to shorten the time required for flowering. In some species vernalization requirements may be replaced in whole or in part by auxin, gibberellin, kinetin, RNA or vitamin E. Note. The low temperature treatment of seeds to induce germination is called, stratification.

Vestiture: Any surface covering such as hairs, scales, etc.

Vexillum: The erect uppermost petal on sweet peas, etc. otherwise known as a standard.

Viable: Able to live; of seeds that have the ability to germinate.

Victorin: A host specific toxin produced by Helminthosporium victoriae utilizable for screening oat cells for resistance to the pathogen.

Vine eye: A wedge-shaped perforated iron nail for driving into a brick wall to support wires up which plants can be trained.

Vine: 1. A grape-vine in Britain and Europe. 2. Any climbing plant in USA.

Vinery: Glasshouse or similar structure. Usually heated for the cultivation of grapes.

Vineyard: An area of land planted with grape vines, usually on a commercial scale for wine-making.

Virescence: The green tinge sometimes present in petals which would not normally contain any green pigment or chlorophyll.

Virgin soil: Soil which has not yet been cultivated.

Virgulate: Shaped like a rod or wand.

Virulent: Highly poisonous or malignant; often describes a fast spreading pathogen that causes serious disease.

Viver: A rootlet or root-fibre.

Viviparous: The strict meaning is producing live young, such as the aphid or greenfly. Also used to describe seed that germinate while still contained in the parent plant, or plantlets produced on the leaves or stems of the parent; such as **Bryophullum daigremontianum*.

Volute: Twinning or twisting; spiraling.

Volva: The membrane protecting many fungi during the early stages of growth.

Vraic: A type of seaweed popular in the Channel Islands for manure and fuel.

V/V: May indicate simple proportions; as 3:1 [v/v]. may indicate present volume in volume; as the number of cm^3 [mls] of constituent in 100 cm^3 [mls] solution. w/m^2 or watt per square metre. A common unit of light measurement. W/V. weight in volume; as the number of grams of constituent in $100cm^3$[mls] solution.

Wall nail: A chisel-shaped nail with a soft lead tang attached to the head. Used for driving into the mortar between bricks so that the lead tang can be twisted around wires or the stems of plants as a support.

Wane: The edge of a plank of wood still carrying the bark, or a plank of wood with a defective edge.

Wardian case: A totally enclosed glass container in which certain plants needing humid conditions such as ferns can be raised. Otherwise known as fern cases. Originally dome-shaped and invented in 1830 by Dr. Nathaniel Bagshaw Ward.

Warren: Formerly an enclosed protected area for the breeding of game; now an area of land where, rabbits live.

Wart: A rounded protuberance on the surface of a plant.

Wasps: Any of a large number of insects of the order Hymenoptera, including the common wasp *Vespa vulgaris*, which is often considered a pest in summer when attracted to ripe fruit and similar sweet substances. In spring and early summer they are very beneficial since they feed their offspring on a variety of other insects.

Water bud: An. underwater bud growing upward to form a new stem on a water plant such as frogbit (*Hydrocharis*).

Water of hydration: The amount of water chemically bound to a substance. This amount may be variable and must be taken into account when solutions of salts are prepared; as in medium preparation.

Water potential: Governs the conduction of plant water from regions of high to low water potential [towards increasing solute concentration] and is a measure of the free energy status of water in a system. The total water potential of system is the sum of the osmotic component [dependent on solute concentration], the matric component [dependent on water-building substances and surface forces] and the pressure component [usually turgor pressure].

Water-soluble powder: A powder formulation which forms a true solution of the active ingredient which dissolved in water.

Wattle: Interlaced thin branches formed into fencing and screens.

Wax: The natural material produced by some plants as a form of protection or to reduce evaporation. It appears in several forms such as the bloom on grapes and plums, a powdery or floury coating on some leaves, or as scales on stems, etc.

Weald: Formerly an area of fairly dense woodland, now a tract of open or slightly wooded country particularly between the North and South Downs of England.

Weed control: The process of limiting weed infestations or killing weeds for aesthetic, economic, public health or other reasons.

Weed cutter: A tool with a narrow horizontal blade attached to a long handle used to server weeds just below ground level. A different design is used to cut weeds in rivers and lakes.

Weed: Any plant growing where it is not wanted. A plant that is useless or of low economic value, especially one growing on cultivated land to the deteriment of the crop.

Weeding fork: A small hand-fork.

Weeping: A tree or shrub of pendulous habit, either natural as in some willows (*Salix*) or induced as in weeping ornamental cherries such as *Prunus serrulata.*

Weevil: 1. A member of the beetle family Curculionidae, usually of small size with elongated head. Most larvae and many beetles cause much damage, particularly to fruit, nuts, bark of trees, etc. 2. Any insect injurious to stored grain.

Wettable powder (Water dispersible powder): A powder formulation which is dispersible in water to form a suspension.

Wettable powder: A finely ground chemical fungicide or pesticide which although not soluble can be suspended in water and applied through a spray if frequently stirred.

Wetting agent: A spreader such as saponin or soft soap which when added to a spraying solution allows the surface of a plant to be covered with a fine mist rather than larger droplets.

Whale hide: A trade naine for bituminized pulpboard made into temporary pots for container-grown plants. No connection with aquatic mammals.

Whin: Another name for gorse.

Whip grafting: A method of grafting by fitting a small tongue and notch cut in the base of the scion into corresponding cuts in the stock.

Whip-and-tongue graft: A graft formed by making a tongue-shaped cut in the scion which fits into a corresponding notch in the stock.

Whip: 1. A long twig or slender, pliant thin branch. 2. A young tree consisting of a single erect stem with no side branches.

White, P.R.: Produced the first successful plant tissue cultures [organ cultures] in the early 1930's. He isolated and grew tomato root tip in a liquid nutrient solution composed or inorganic salts, sucrose and yeast extract [1934]. He showed that yeast could be replaced with the B vitamins, thiamine, pyridoxine and niacin [1937]. He wrote A Handbook of Plant Tissue Culture [1943], The cultivation of Animal and Plant Cells [1954]. In 1963 he published a low salt nutrient culture medium widely used in plant tissue culture.

Whorl: A circle of three or more flowers or leaves at one node, arranged as the spokes of a wheel and springing from stem or ax is at the same level. .

Wick: A filter paper, chromatography paper or a strip of some other material used to support plant tissue above a liquid medium and to transport the water and nutrients to the tissue.

Wicker: A small flexible osier or twig, used for making wicker work or basketry.

Wicket: A small gate often adjacent to a larger one.

Wigwam: Three or more canes or poles arranged in a conical shape and tied at the top to form a pyramid as a support for climbing plants such as sweet peas runner beans. etc. during the winter months, not planted.

Wild type: 1. The genotype or phenotype of an organism predominating in the control [standard or wild] population, in its natural element. 2. A specific gene predominant in this population.

Wilting point: The time at which the water content of a soil is reduced to a level at which permanent wilting occurs.

Win: To dry by exposure to the wind.

Wind pollination: The transfer of male pollen to female flowers by the action of wind or air currents, as in hazel and many other catkin bearing plants.

Wind break: A group or row of plants, trees or material fixed to stakes, specifically arranged to break the force of a strong wind. A permeable windbreak will give protection for a distance of about 10ft for each foot of height; whereas a solid wall or fence is effective for about twice its height but creates violent turbulence beyond.

Windfall: Fruit which drops to the ground naturally, either when ripe or diseased.

Windlestraw: A thin dry stalk of grass; any of the long-stalked species of grass.

Wing: 1. One of the pair of lateral petals of papilionaceous flowers such as the sweet pea. 2. A dry flat appendage of a seed to assist its distribution by wind or air currents, such as sycamore.

Winnow: To separate the chaff from the grain by wind action; to sift or separate by wind action.

Winter fallow: Agricultural term for land left vacant during the winter months, not planted.

Winter spores: The special type of spore produced by fungi in autumn which are adapted to withstanding winter conditions.

Wire rake: Same as lawn rake or springbok rake; one made of fanned out sprung wires bent at the ends.

Wireworms: The larvae of click beetles (*Agriotes lineatus*) which live in the soil The grubs are serious pests and attack the roots of many crops including tubers, bulbs and corms, They are particularly prevalent in old pasture and in soils which have been recently reclaimed from grassland.

Wisp: 1. A small bundle of hay, straw, etc. as used in the making of com-dollies, etc. 2. A small hand-broom or brush, originally made from a small bundle of straw.

Witch's Broom: I. A tuft of underdeveloped branches produced usually on woody plants and trees; if cuttings are taken from them they will remain dwarf. Often caused by fungal attack. 2, Bundle of twigs tied to a rustic handle used for sweeping leaves, etc. from lawns. A besom.

Withe: A flexible twig usually of willow; a handle for a basket or tool made from a number of withes twisted together.

Withwind: A climbing plant such as honeysuckle or bindweed.

Withy: Flexible shoot of willow or osier grown specially for use in the manufacture of baskets and for tying bundles of canes or rods.

Wold: Formerly a wooded area, now usually upland country.

Wood ash: The substance obtained by burning wood and similar materials which contains potash, phosphate, iron, magnesium and manganese in varying amounts. It is very soluble in water, so should be added to the soil immediately or stored in a dry place.

Wood: 1. A group of trees growing fairly close together but not so densely as a forest and not so extensive in area. 2. The hard part of the substance of trees and shrubs, xylem.

Woodlice: Arthropods which inhabit dark moist situations such as decaying vegetation, bark, etc. They will sometimes eat seedlings but, having soft. mouthparts, in, general present no threat to the gardener.

Woody plant medium (WPM): A modified MS [1962] medium developed for woody plant species by G. Lloyd and B. McCown [1980]. It has less nitrogen [both ammonium and nitrate], sodium, potassium and chloride than does MS [1962] medium. It has been increasingly used for the commercial propagation of ornamental trees and shrubs.

Woody: Of a plant with with lignified stems and branches, usually somewhat lacking in foliage.

Woolly aphid: A type of aphid which secretes and surrounds itself with a waxy, cottony substance which is resistant to water. They attack several kinds of; tree, including pear and apple several kinds of tree, including pear and apple. Also called American blight.

Woolly: Having long, soft matted hairs; pubescent; downy.

Worm: 1. Slender burrowing elongate invertebrate animal of the genus *Lumbricus* with reddish-brown body. The gardener's greatest asset, known to the Victorians as nature's ploughman. 2. The general name for a small maggot or grub, especially in USA.

Wrack: Seaweed cast ashore or growing where it is exposed by the tide; any of the Fucaceae, the bladder-wrack family of seaweeds.

Wrenching: A method of root-pruning carried out in commercial nurseries to make later transplanting easier, or to prevent the development of a tap-root.

Xanthein: The yellow colouring matter present in trees e.g. *Xanthoria parietina,* the lichen often seen on walls and roots.

Xanthic: Yellow, or the yellow pigment in flowers, often combined with carotene in yellow flowers and fruit.

Xanthomonas: Genus of bacterium that can cause leaf spots on plants

Xanthophyll: The dark brown or yellow compound in plants forming the colouring matter of autumn foliage.

Xenogamy: Cross-fertilization; fertilization between flowers on different plants.

Xenograft: A graft from a member of a different species.

Xeromorphic: Plants, or parts of plants, protected against excessive loss of moisture by hairs, thick cuticles, or similar structural characteristics.

Xerophyte: A plant adapted to inhabit places where fresh water is scanty, or where water uptake is difficult, owing to an excess of salts, e.g, on a seashore. Tamarisk (*Tamarix anglica*) is an example.

Xerophytic: A plant capable of survival under conditions of extreme drought, or one adapted to arid conditions, e.g. cacti.

Xiphophyllus: A type of plant having sword-shaped leaves, e.g. iris.

Xylem: Woody tissue in plants usually consisting of vessels, fibres, etc., all with lignified walls, for the conducting of aqueous solutions and for support.

Xylocarp: A hind, woody fruit such as the coconut.

Xylogenesis: The formulation of xylem in a vascular bundle.

Xylogenous: Of organisms which grow on wood; some fungi are xylogenous.

Xylology: The study of the composition and structure of wood.

Xylophagous: Insects which feed on, or bore into wood, e.g. the death watch beetle.

Xylose [$C_5H_{10}O_5$ mw 150:1]: An aldopentose sugar [wood sugar] present in the woody tissues of many plants as polymeric xylan. It is occasionally used a carbohydrate source in plant tissue culture media.

Xystus: A tree-planted walk, an avenue between trees.

Yard: 1. A straight thin branch. 2. USA term for a garden, especially one at the rear of the house; backyard. 3. Unit of length equal to 3 feet 90 cm)

Year-round flowering: Control of day length and temperature to produce flowering plants throughout the year. Often used to describe azalea and chrysanthemum programs.

Yield test: Productivity assessment

Yield: The net recoverable amount of the desired product from a reaction or process. This may be expressed in terms of the percentage yield of the ratio of final product to starting material.

Zap: To hit, strike or kill swiftly, as in the swatting of a fly. Probably of USA origination.

Zeatin or 2-methyl-4-[1H-purin-6-ylamino]-2-buten 1-ol [Z, $C_{10}H_{13}N_5O$ mw 215:21]: A natural cytokinin first isolated from corn. It is sometimes included in plant tissue culture media. It dissolves in acid [HCl ca 1 M].

Zeatin riboside [mw 351:4]: A naturally occurring cytokinin sometimes used in plant tissue culture media.

Zebra: Any plant or tree having stripes reminiscent of a zebra's. The hard and striped wood of Guaiacum is called zebra-wood.

Zephiran: The brand name for a disinfecting agent for plant material containing benzalkonium chloride.

Zone lines: Thin black or dark brown lines in decayed wood, usually caused by fungi.

Zoning: 1. Leaf-markings, usually circular and darker than the main leaf area but can be lighter or differently coloured. 2. Climatic zones detailed in USA and other large countries to help assess the hardiness of plants in a particular part of the country.

Zoophyte: Any plant thought to resemble an animal.

Zoospore: A spermatozoid, an asexual reproductive cell. A spore occurring in certain fungi algae, etc. having power of locomotion. Some rusts are reproduced by zoospores.

Zygomycetes: A group of fungi, moulds etc. which produce zygospores.

Zygophyte: A plant which reproduces by means of zygospores.

Zygospores: A spore produced by the union of buds from two adjacent hyphae; the method by which some fungi and algae multiply.

Zymase: Any group of enzymes inducing fermentation.

Zymolysis: The action of enzymes.

❑❑❑

4 Soil Science

A horizon: The surface horizon of a mineral soil having maximum organic matter accumulation, maximum biological activity and/ or eluviation of materials such as iron and aluminum oxides and silicate clays.

ABC soil: A soil with a distinctly developed profile, including A, B, and C horizons.

Absorption, active: Movement of ions and water into the plant root as a result of metabolic processes by the root, frequently against an activity gradient.

Absorption passive: Movement of ions and Water into the plant root as a result of diffusion along a gradient.

AC soil: A soil having a profile containing only A and C horizons with no clearly developed B horizon.

Accelerated erosion: erosion much more rapid than normal, natural, geological ereosion; primarlily as a result of the influence of the activities of humans or in some cases, of animals.

Acid rain: Atmospheric precipitation with pH values less than about 5.6, the acidity bring due to inorganic acids such as nitric and sulfuric that are formed when oxides of nitrogen and sulfur are emitted into the atmosphere.

Acid soil: A soil with a pH value <7.0 usually applied to surface layer or root zone, but may be used to characterize any horizon. *See also* reaction, soil.

Acidity, active: The activity of hydrogen ion in the aqueous phase of a soil. It is measured and expressed as a pH value.

Acidity, exchange (reserve): Exchangeable hydrogen, aluminum, and other cations that, when removed from the exchange complex, are sources of soil solution hydrogen ions. This acidity must be neutralized to bring an acid soil to neutrality or to some predetermined higher pH. Usually expressed in centimoles per kilogram (cmol/kg) of soil.

Actinomycetes: A group of organisms intermediate between the bacteria and the true fungi that usually produce a characteristic branched mycelium. Includes many, but not all, organisms belonging to the order of Actinomycetales.

Activated sludge: Sludge that has been aerated and subjected to bacterial action.

Adhesion: Molecular attraction that holds the surfaces of two substances (e.g., water and sand particles) in contact.

Adsorption complex: The group of substances in soil capable of adsorbing other materials. Colloidal particles account for most of this adsorption.

Adsorption: The attraction of ions or compounds to the surface of a solid. Soil colloids adsorb large amounts of ions and water.

Aerate: To impregnate with gas, usually air

Aeration soi1: The process by which air in the soil is replaced by air from the atmosphere. In a well-aerated soil, the soil air is very similar in composition to the atmosphere above the soil. Poorly aerated soils usually contain more carbon dioxide and correspondingly less oxygen than the atmosphere above the soil.

Aerobic: 1. Having molecular oxygen as a part of the environment. 2. Growing only in the presence of molecular oxygen, as aerobic organisms. 3. Occurring only in the presence of molecular oxygen (said of certain chemical or biochemical processes, such as aerobic decomposition).

Aggregate (soil): Many soil particles held in a single mass or cluster such as a clod, crumb, block, or prism.

Agronomy: A specialization of agriculture concerned with the theory and practice of field-crop production and soil management. The scientific management of land.

Air porosity: The proportion of the bulk volume of soil that is filled with air at any given time or under a given condition, such as a specified moisture tension. Usually the large pores: that is, those drained by a tension of less than approximately 100 cm of water.

Air-dry: 1. The state of dryness (of a soil) at equilibrium with the moisture content in the surrounding atmosphere. The actual moisture content will depend upon the relative humidity and the temperature of the surrounding atmosphere. 2. To allow to reach equilibrium in moisture content with the surrounding atmosphere.

Albic horizon: A mineral soil horizon from which clay and free iron oxides have been removed or in which the oxides have been segregated to the extent that the color of the horizon is determined primarily by the color of the primary sand and silt particles rather than by coatings on these particles.

Alfisols: Soils with gray to brown surface horizons, medium to high supply of bases, and B horizons of illuvial clay accumulation. These soils form mostly under forest or savanna vegetation in climates with slight to pronounced seasonal moisture deficit.

Alkali soil: 1. A soil with a high degree of alkalinity ($pH \geq 8.5$) or with a high exchangeable sodium content (15% or more of the exchange capacity), or both. 2. A soil that contains sufficient alkali (sodium) to interfere with the growth of most crop plants.

Alkaline soil: Any soil that has $pH > 7$. Usually applied to surface layer or root zone but may be used to characterize any horizon or a sample thereof.

Alluvial soil: (Obsolete) A soil developing from recently deposited alluvium and exhibiting essentially no horizon development or modification of the recently deposited materials.

Alluvium: A general term for all detrital material deposited or in transit by streams, including gravel, sand, silt, clay, and all variations and mixtures of these. Unless otherwise noted, alluvium is unconsolidated.

Alpine Meadow soils: A great soil group of the intrazonal order, comprised of dark soils of grassy meadows at altitudes above the timberline.

Alumino-silicates: Compounds containing aluminum, silicon, and oxygen as main constituents. An example is microcline, $KAlSi_3O_8$.

Amendment, soil: Any substance such as lime, sulfur, gypsum, and sawdust used to alter the properties of a soil, generally to make it more productive. Strictly speaking, fertilizers are soil amendments, but the term is used most commonly for materials other than fertilizers.

Amino acids: Nitrogen-containing organic acids that couple together to form proteins. Each acid molecule contains one or more amino groups ($-NH_2$) and at least one carboxyl group ($-COOH$). In addition, some amino acids contain sulfur.

Ammonification: The biochemical process whereby ammoniacal nitrogen is released from nitrogen-containing organic compounds.

Ammonium fixation: The adsorption of ammonium ions by the mineral or organic fractions of the soil in forms that are relatively insoluble in water and relatively non-exchangeable.

Anaerobic: 1. Without molecular oxygen. 2. Living or functioning in the absence of air or free oxygen.

Ando soils (Obsolete): A zonal group of dark- colored soils high in organic matter developed in volcanic ash deposits.

Anion exchange capacity: The sum total of exchangeable anions that a soil can absorb. Expressed as centimoles per kilogram (cmol/kg) of soil (or of other adsorbing material such as clay).

Anion: Negatively charged ion; ion that during electrolysis is attracted to the anode.

Anthropic epipedon: A surface layer of mineral soil that has the same requirements as the mollic epipedon but that has more than 250 ppm of P_2O_5 soluble in 1% citric acid, or is dry more than 10 months (cumulative) during the period when not irrigated. The anthropic epipedon forms under long-continued cultivation and fertilization.

Antibiotic: A substance produced by one species of organism that, in low concentrations, will kill or inhibit growth of certain other organisms.

Ap: The surface layer of a soil disturbed by cultivation or pasturing.

Apatite: A naturally occurring complex calcium phosphate that is the original source of most of the phosphate fertilizers. Formulas such as $3Ca_3(PO_4)_2 CaF_2$ illustrate the complex compounds that make up apatite.

Argillic horizon: A mineral soil horizon characterized by the illuvial accumulation of layer-lattice silicate clays.

Aric horizon: A mineral soil horizon in which clay, silt, and humus derived from an overlying cultivated and fertilized layer have accumulated. The wormholes and illuvial clay, silt, and humus, occupy at least 5% of the horizon by volume.

Aridisols: Soils of dry areas. They have pedogenic horizons, low in organic matter that is never moist as long as 3 consecutive months. They have an ochric epipedon and one or more of the following diagnostic horizons: argillic, natric, cambic, calcic, petrocalcic, gypsic, salic, or a duripan.

Atterberg limits. Atterberg limits are measured for soil materials passing through serve the No. 40.

Autotroph: An organism capable of utilizing carbon dioxide or carbonates as the sole source of carbon and obtaining energy for life processes from the oxidation of inorganic elements or compounds such as iron, sulfur, hydrogen. ammonium and nitrites, or from radiant energy. *Contrast* with heterotroph.

Available nutrient: That portion of any element or compound in the soil that can be readily absorbed and assimilated by growing plants. "Available" should not be confused with "exchangeable".

Available water: The portion of water in a soil that can be readily absorbed by plant roots. Considered that water held in the soil against a pressure of up to approximately 15 bars.

Azonal soils (Obsolete): Soils without distinct genetic horizons.

B horizon: A soil horizon usually beneath the A that is characterized by one or more of the following: (a) a concentration of silicate clays, iron and aluminum oxides, and humus, alone or in combination; (b) a blocky or prismatic structure, and (c) coatings of iron and aluminum oxides that give darker, stronger or redder color.

Bar: A unit of pressure equal to one million dynes per square centimeter (10^6 dynes/cm^2).

Base saturation percentage: The extent to which the adsorption complex of a soil is saturated with exchangeable cations other than hydrogen and aluminum. It is expressed as a percentage of the total cation exchange capacity.

BC soil: A soil profile with B and C horizons but with little or no A horizon. Most BC soils have lost their A horizons by erosion.

Bedding (Engineering): Arranging the surface of fields by plowing and grading into a series of elevated beds separated by shallow depressions or ditches for drainage.

Bedrock: The solid rock underlying soils and the regolith in depths ranging from zero (where exposed by erosion) to several hundred feet.

Bench terrace: An embankment constructed across sloping fields with a steep drop on the downslope side.

Biodegradable: A material subject to degradation by biochemical processes.

Biological nitrogen fixation: Occurs at ordinary temperatures and pressures It is commonly carried out by certain bacteria, algae and actinomycetes which may or may not be associated with higher plants.

Biomass: The amount of living matter in a given area.

Bleicherde: The light-colored, leached A_2 horizon of Spodosols.

Blown-out land: Aras from which all or almost all of the soil and soil material has been removed by wind erosion. Usually unfit for crop production. A miscellaneous land type.

Bog soil: A great soil group of the intrazonal order and hydromorphic suborder. Includes muck and peat.

Border-strip: The water is applied at the upper end of a strip with earth borders to confine the water to the strip.

Breccia: A rock composed of coarse angular fragments cemented together.

Broad-base terrace: A low embankment with such gentle slopes that it can be farmed, constructed across sloping fields to reduce erosion and runoff.

Brown Earths: Soils with a mull horizon but having no horizon of accumulation of clay or sesquioxides. (Generally used as a synonym for "Brown Forest soils" but sometimes for similar soils acid in reaction).

Brown Forest soils: A great soil group of the intrazonal order and calcimorphic suborder, formed on calcium-rich parent materials under deciduous forest, and possessing a high base status but lacking a pronounced illuvial horizon.

Brown Podzollic soils: A zonal great soil group similar to Podzols but lacking the distinct A2 horizon characteristic of the Podzol group.

Brown soils: A great soil group of the temperate to cool arid regions, composed of soils with a brown surface and a light-colored transitional subsurface horizon over calcium carbonate accumulation. They develop under short grasses.

Brunizem: Synonymous with *Prairie soils.*

Buffer compounds, soil: The clay, organic matter, and compounds such as carbonates and phosphates that enable the soil to resist appreciable change in pH.

Bulk blended fertilizers: Solid fertilizer materials blended together in small blending plants delivered to the farm in bulk, usually spread directly on the fields by truck or other special applicator

Bulk density, soil: The mass of dry soil per unit of bulk volume, including the air space. The bulk volume is determined before drying to constant weight at 105°C.

Buried soil: Soil covered by an alluvial, loessal, or other deposit, usually to a depth greater than the thickness of the solum. C horizon. A mineral horizon generally beneath the solum that is relatively unaffected by biological activity and pedogenesis and is lacking properties diagnostic of an A or B horizon. It may or may not be like the material from which the A and B have formed.

Calcareous soil: Soil containing sufficient calcium carbonate (often with magnesium carbonate) to effervesce visibly when treated with cold 0.1 N hydrochloric acid.

Caliche: A layer near the surface, more or less cemented by secondary carbonates of calcium or magnesium precipitated from the soil solution. It may occur as a soft thin soil horizon, as a hard thick bed just beneath the solum, or as a surface layer exposed by erosion.

Callcic horizon: A mineral soil horizon of secondary carbonate enrichment that is more than 15 cm thick, has a calcium carbonate equivalent of more than 15%, and has at least 5% more calcium carbonate equivalent than the underlying C horizon.

Cambic horizon: A mineral soil horizon that has a texture of loamy very fine sand or finer, contains some weatherable minerals, and is characterized by the alteration or removal of mineral material. The cambic horizon lacks cementation or induration and has too few evidences of illuviation to meet the requirements of the argillic or spodic horizon.

Capillary porosity: The small pores, or the bulk volume of small pores, which hold water in soils against a tension usually >60 cm of water.

Capillary water (Obsolete): The water held in the "capillary' or small pores of a soil, usually with a tension >60 cm of water.

Carbon cycle: The sequence of transformations whereby carbon dioxide is fixed in living organisms by photosynthesis or by chemosynthesis, liberated by respiration and by the death and decomposition of the fixing organism, used by heterotrophic species, and ultimately returned to its original state.

Carbon/nitrogen ratio: The ratio of the weight of organic carbon (C) to the weight of total nitrogen (N] in a soil or in organic material.

Cat clays: Wet clay soils high in reduced forms of sulfur that upon being drained become extremely acid due to the oxidation of the sulfur compounds and the formation of sulfuric acid.

Catena: A sequence of soils of about the same age, derived from similar parent material, and occurring under similar climatic conditions, but having different characteristics due to variation in *relief* and in *drainage.*

Cation exchange capacity: The sum total of exchangeable cations that a soil can adsorb. Sometimes called 'total-exchange capacity," "base-exchange capacity, or "cation-adsorption capacity.' Expressed in centimoles per kilogram (cmol/kg) of soil (or of other adsorbing material such as clay.

Cation exchange: The interchange between a cation in solution and another cation on the surface of any surface-active material such as clay or organic matter.

Cation: A positively charged ion.

Cemented: Indurated; having a hard, brittle consistency because the particles are held together by cementing substances such as humus, calcium carbonate, or the oxides of silicon, iron, and aluminum.

Center pivot: Automated sprinkler irrigation achieved by automatically rotating the sprinkler pipe or boom, supplying water to the sprinkler heads or nozzles, as a radius from the center of the field to be irrigated.

Channery: Thin, flat fragments of limestone, sandstone, or schist up to 6 in. (15 cm) in major diameter.

Chart: A structureless form of silica, closely related to flint that breaks into angular fragments.

Check-basin: The water is applied rapidly to relatively level plots surrounded by levees. The basin is a small check.

Chemical nitrogen fixation: Takes place at high temperatures and pressures in manufacturing plants; produces ammonia, which is used to manufacture most fertilizers.

Chernozem: A zonal great soil group consisting of soils with a thick nearly black or black, organic matter-rich A horizon high in exchangeable calcium, underlain by a lighter colored transitional horizon above a zone of calcium carbonate accumulation; occurs in a cool subhumid climate under a vegetation of tall and mid-grass prairie.

Chessnut soil: A zonal great soil group consisting of soils with a moderately thick, dark-brown A horizon over a lighter colored horizon that is aove a zone of calcium carbonate accumulation. They develop under mixed tall and short grasses in a temperate to cool and subhumid semiarid climate. [1949 classification system].

Chroma: The relative purity, strength or saturation of a colour directly related to the dominance of the determining wavelength of the light and listed to grayness; one of the three variables of color.

Class, soil: A group of soils having a definite range in a particular property such as acidity, degree of slope, texture, structure, land-use Capability, degree of erosion, or drainage.

Clastic: Composed of broken fragments of rocks and minerals.

Clay mineral: Naturally occurring inorganic material (usually crystalline] found in soils and other earthy deposits, the particles being of clay size, that is, <0.002 mm in diameter.

Clay: 1. A soil separate consisting of particles <0.002 mm in equivalent diameter. 2. A soil textural class containing >40% clay, <45% sand, and <40% silt.

Claypan: A dense, compact, slowly permeable layer in the subsoil having a much higher clay content than the overlying material, from which it is separated by a sharply defined boundary. Claypans are usually hard when dry, and plastic and sticky when wet.

Clod: A compact, coherent mass of soil produced artificially, usually by the activity of man by plowing, digging, and so on, especially when these operations are performed on soils that are either too wet or too dry for normal tillage operations.

Coarse texture: The texture exhibited by sands, loamy sands, and sandy loams except very fine sandy loam.

Cobblestone: Rounded or partially rounded rock or mineral fragments 3-10 in. in diameter.

Cohesion: Holding together: force holding a solid or liquid together, owing to attraction between like molecules, Decreases with rise in temperature.

Colloid, soil (Greek, glue-like): Organic and inorganic matter with very small particle size and a correspondingly large surface area per unit of mass.

Colluvium: A deposit of rock fragments and soil material accumulated at the base of steep slopes as a result of gravitational action.

Compost: Organic residues, or a mixture of organic residues and soil, that have been piled, moistened, and allowed to undergo biological decomposition. Mineral fertilizers are sometimes added. Often called "artificial manure' or "synthetic manure' if produced primarily from plant residues.

Consistence: The combination of properties of soil material, that determine its resistance to crushing and its ability to be molded or changed in shape. Such terms as loose, friable, firm, soft, plastic, and sticky describe soil consistence.

Consumptive use: The water used by plants in transpiration and growth, plus water vapor loss from adjacent soil or snow, or from intercepted precipitation in any specified time. Usually expressed as equivalent depth of free water per unit of time.

Contour: An imaginary line connecting points of equal elevation on the surface of the soil. A contour terrace is laid out on a sloping soil at right angles to the direction of the slope and nearly level throughout its course.

Contour strip cropping: Layout of crops in comparatively narrow strips in which the farming operations are performed approximately on the contour. Usually strips of grass, close-growing crops, or fallow are alternated with those in cultivated crops.

Convection: The transfer of heat through a gas or solution because of molecular movement.

Corrugation: The water is applied to small, closely-spaced furrows, frequently in grain and forage craps, to confine the flow of irrigation water to one direction.

Cover crop: A close-growing crop grown primarily for the purpose of protecting and improving soil between periods of regular crop production or between trees and vines in orchards and vineyards.

Creep: Slow mass movement of soil and soil material down relatively steep slopes primarily under the influence of gravity, but facilitated by saturation with water and by alternate freezing and thawing.

Crop rotation: A planned sequence of crops growing in a regularly recurring succession on the same area of land, as contrasted to continuous culture of one crop or growing different crops in haphazard order

Crumb: A soft, porous, more or less rounded natural unit of structure from 1 to 5 mm in diameter. *See also* soil structure types.

Crushing strength: The force required to crush a mass of dry soil or, conversely, the resistance of the dry soil mass to crushing. Expressed in units of force per unit area (pressure).

Crust: A surface layer on soils, ranging in thickness from a few millimeters to perhaps as much as 3 cm, that is much more compact, hard, and brittle, when dry than the material immediately beneath it.

Crystal: A homogeneous inorganic substance of definite chemical composition bounded by plane surfaces that form definite angles with each other, thus giving the substance a regular geometrical form.

Crystalline rock: A rock consisting of various minerals that have crystallized in place from magma.

Cultivation: A tillage operation used in preparing land for seeding or transplanting or later for weed control and for loosening the soil.

Deciduous plant: A plant that sheds all its leaves every year at a certain season.

Deflocculate: 1. To separate the individual components of compound particles by chemical and/or physical means. 2. To cause the particles of the *disperse phase* of a colloidal system to become suspended in the *dispersion medium.*

Denitrification: The biochemical reduction of nitrate or nitrite to gaseous nitrogen, either as molecular nitrogen or as an oxide of nitrogen.

Desalinization: Removal of salts from saline soil, usually by leaching.

Desert crust: A hard layer, containing calcium carbonate, gypsum, or other binding material, exposed at the surface in desert regions.

Desert soil: A zonal great soil group consisting toils with a very thin, light-colored surface horizon which may be vesicular and is ordinarily underlain by calcareous material; formed in arid regions under sparse shrub vegetation.

Desorption: The removal of sorbed material from surfaces.

Diagnostic horizons (As used in *Soil Taxonomy)***:** Combinations of specific soil characteristics that are indicative of certain classes of soils. Those that occur at the soil surface are called epipedons; those below the surface, diagnostic subsurface horizons.

Diagnostic surface horizons: The following diagnostic surface horizons are used in *Soil Taxonomy* and are called epipedons.

Diatom: Algae having siliceous cell walls that persist as a skeleton after death; any of the microscopic unicellular or colonial algae constituting the class Bacillariaceae. They occur abundantly in fresh and salt waters and their remains are widely distributed in soils.

Diatomaceous earth: A geologic deposit of fine, grayish, siliceous material composed chiefly or wholly of the remains of diatoms. It may occur as a powder or as a porous, rigid material.

Diffusion: The transport of matter as a result of the movement of the constituent particles. The intermingling of two gases or liquids in contact with each other takes place by diffusion.

Disintegration: The breakdown of rock and mineral particles into smaller particles by physical forces such as frost action.

Disperse: 1. To break up compound particles, such as aggregates, into the individual component particles. 2. To distribute or suspend fine particles, such as clay, in or throughout a dispersion medium, such as water.

Diversion dam: A structure or barrier built to divert part or all of the water of a stream to a different course.

Drain, to: 1. To provide channels, such as open ditches or drain tile, so th't excess water can be removed by surface or by internal flow. 2. To lose water (from the soil) by percolation.

Drainage, soil: The frequency and duration of periods when the soil is free of saturation with water.

Drift: Material of any sort deposited by geological processes in one place after having been removed from another. Glacial drift includes material moved by the glaciers and by the streams and lakes associated with them.

Drip: A planned irrigation system where all necessary facilities have been installed for the efficient application of water directly to the root zone of plants by means of applicators (orifices, emitters, porous tubing, perforated pipe, etc.) operated under low pressure. The applicators may be placed on or below the surface of the ground.

Drumlin: Long, smooth cigar-shaped low hills of glacial till, with their long axes parallel to the direction of ice movement.

Dryland farming: The practice of crop production in tow-rainfall areas without irrigation.

Duff: The matted, partly decomposed organic surface layer of forest soils.

Duripan: A mineral soil horizon that is cemented by silica, to the point that air-dry fragments will not slake in water or Hcl. A dun-pan may also have accessory cement such as iron oxide or calcium carbonate.

Dust mulch: A loose, finely granular, or powdery condition on the surface of the soil, usually produced by shallow cultivation.

Ectotrophic mycorrhiza (ectomycorrhiza): A mycorrhizal association in which the fungal hyphae form a compact mantle on the surface of the roots and extend into the surrounding soil. Mycelial strands extend inward between corti but not into these cells. Associated primarily with certain trees. See also endotrophic mycorrhiza.

Electrokinetic potential: In a colloidal system, the difference in potential between the immovable layer attached to the surface of the dispersed phase and the dispersion medium.

Eluviation: The removal of soil material in suspension (or in solution) from a layer or layer of the soil. (Usually, the loss of material in solution is described by the term "leaching") seed also leaching.

Endotrophic mycorrhiza (endomycorrhiza): A mycorrhizal association in which fungal hyphae penetrate directly into root hairs, other epidermal cells, and occasionally into cortical cells, individual threads extend from the root surface outward into the surrounding soil.

Endotrophic: Nourished or receiving nourishment from within, as fungi or their hyphae receiving nourishment from plant roots in a mycorrhizal association.

Entisols: Soils that have no diagnostic pedogenic horizons. They may be found in virtually any climate on very recent geomorphic surfaces.

Epipedon: A diagonostic surface horizon that includes the upper part of the soil that is darkened by organic matter, or the upper eluvial horizons or both (Soil Taxonomy).

Erosion: 1. The wearing away of the land surface by running water, wind, ice, or other geological agents, including such processes as gravitational creep. 2. Detachment and movement of soil or rock by water, wind, ice, or gravity. The following terms are used to describe different types of water erosion.

Erosion: The erosion process whereby water accumulates in narrow channels and, over short periods, removes the soil from this narrow area to considerable depths, ranging from 1-2 feet to as much as 75-100 feet.

Essential element: A chemical element required for the normal growth of plants.

Eutrophic: Having concentrations of nutrients optimal (or nearly so) for plant or animal growth. (Said of nutrient solutions or of soil solutions).

Eutrophication: A means of aging of lakes whereby aquatic plants are abundant and waters are deficient in oxygen. The process is usually accelerated by enrichment of waters with surface runoff containing nitrogen and phosphorus.

Evapotranspiration: The combined loss of water from a given area, and during a specified period of time, by evaporation from the soil surface and by transpiration from plants.

Exchange acidity: The titratable hydrogen and aluminum that can be replaced from the adsorption complex by a neutral salt solution. Usually expressed as centimoles per kilogram (cmol/kg) of soil.

Exchange capacity: The total Ionic charge of the adsorption complex active in the adsorption of ions. *See also* anion exchange capacity; cation exchange capacity.

Exchangeable-sodium percentage: The extent to which the adsorption complex of a soil is occupied by sodium. It is expressed as follows.

$$\text{ESP} = \frac{\text{Exchangeable sodium (cmol/kg soil)}}{\text{Cation exchange capacity (cmol/kg soil)}} \times 100$$

Fallow: Cropland left ideal in order to restore productivity, mainly through accumulation of water, nutrients, or both. Summer fallow is a common stage before cereal grain in regions of limited rainfall. The soil is kept free of weeds and other vegetation, thereby conserving nutrients and water for the next year's crop.

Family, soil: In soil classification, one of the categories intermediate between the great group and the soil series. Families are defined largely on the basis of physical and mineralogical properties of importance to plant growth.

Fertility, soil: The status of a soil with respect to the amount and availability to plants of elements necessary for plant growth.

Fertilizer: Any organic or inorganic material of natural or synthetic origin added to a soil to supply certain elements essential to the growth of plants. The major types of fertilizers include:

Fibric materials: The least decomposed of all the organic soil materials, containing very high amounts of fiber that are well preserved and readily identifiable as to botanical origin.

Field capacity (field moisture capacity): The percentage of water remaining in a soil two or three days after having been saturated and after free drainage has practically ceased.

Fine, grained mica: A silicate clay having a 2:1-type lattice structure with much of the silicon in the tetrahedral sheet having been replaced by aluminum and with considerable interlayer potassium which binds the layers together and prevents interlayer expansion swelling and limits interlayer cation exchange capacity.

Fine texture: Consisting of or containing large quantities of the fine fractions, particularly of silt and clay. (Includes clay loam, sandy. clay loam, silty clay loam, sandy clay, silty clay, and clay textural classes)

First bottom: The normal floodplain of a stream.

Fixation: 1. For other than elemental nitrogen: the process or processes in a soil by which certain chemical elements essential for plant growth are converted from a soluble or exchangeable form to a much less soluble or to a nonexchangeable form; for example, potassium, ammonium and phosphate "fixation." 2. For elemental nitrogen: process by which gaseous elemental nitrogen is chemically combined with hydrogen to form ammonia.

Flagstone: A relatively thin rock or mineral fragment 15-38 cm in length commonly composed of shale, slate, limestone, or sandstone.

Flocculate: To aggregate or clump together individual, tiny soil particles, especially fine clay, into small clumps or floccules. Opposite of *deflocculate* or *disperse.*

Flooding: The water is released from field ditches and allowed to flood over the land.

Flood plain: The land bordering a stream, built up of sediments from overflow of the stream and subject to inundation when the stream is at flood stage. Sometimes called bottomland.

Flora: The sum total of the kinds of plants in a area at one time.

Fluor apatite: A member of the apatite group of minerals containing fluorine. Most common mineral in rock phosphate.

Foliar diagnosis: An estimation of mineral nutrient deficiencies (excesses) of plants based on examination of the chemical composition of selected plant parts, and the color and growth characteristics of the foliage of the plants.

Fragipan: Dense and brittle pan or subsurface layer in soils that owe their hardness mainly to extreme density or compactness rather than high clay content or cementation. Removed fragments are friable, but the material in place is so dense that roots penetrate and water moves through it very slowly

Friable: A soil consistency term pertaining to the ease of crumbling of soils.

Fritted micronutrients: Sintered silicates having total guaranteed analyses of micronutrients with controlled (relatively slow) release characteristics.

Fulvic acid: A term of varied usage but usually referring to the mixture of organic substances remaining in solution upon acidification of a dilute alkali extract from the soil.

Fungi: Simple plants that tack a photosynthetic pigment. The individual cells have a nucleus surrounded by a membrane, and they may be linked together in long filaments called *hyphae,* which may grow together to form a visible body.

Furrow: The water is applied to row crops in ditches made by tillage implements.

Genesis, soil: The mode of origin of the soil, with special reference to the processes responsible for the development of the solum, or true soil, from the unconsolidated parent material.

Gilgai: The microrelief of soils produced by expansion and contraction with changes in moisture. Found in soils that contain large amounts of clay that swells and shrinks considerably with wetting and drying. Usually a succession of microbasins and microknolls in nearly level areas or of microvalleys and microridges parallel to the direction of the slope.

Glacial drift: Rock debris that has been transported by glaciers and deposited, either directly from the ice or from the melt-water. The debris may or may not be heterogeneous.

Glaciofluvial deposits: Material moved by glaciers and subsequently sorted and deposited by streams flowing from the melting ice. The deposits are stratified and may occur in the form of outwash plains, deltas, kames, eskers, and kame terraces.

Gley soil: (Obsolete) Soil developed under conditions of poor drainage resulting in reduction of iron and other elements and in gray colors and mottles.

Granular structure: Soil structure in which the individual grains are grouped into spherical aggregates with indistinct sides. Highly porous granules are commonly called crumbs. **A** well granulated soil has the best structure for most ordinary crop plants. See also soil structure types.

Granulated fertilizers: Fertilizers that are present in the form of rather stable granules of uniform size, which facilitate ease of handling the materials and reduce undesirable dusts.

Gravitational potential: That portion of the total soil water potential due to differences in elevation of the reference pool of pure water and that of the soil water. Since the soil water elevation is usually chosen to be higher than that of the reference pool, the gravitational potential is usually positive.

Gravitational water: Water that moves into through, or out of the soil under the influence of gravity.

Gray Wooded soils: A zonal great soil group consisting of soils with a thin Al horizon over a light-colored, bleached (A_2) horizon underlain by a B horizon containing a high percentage of bases and an appreciable quantity of illuviated silicate clay. These soils occur in subhumid to semiarid, cool climatic regions under coniferous, deciduous, or mixed forest cover.

Gray-Brown Podzolic soil: A zonal great soil group consisting of soils with a thin, moderately dark Al horizon and with a grayish brown A_2 horizon underlain by a B horizon containing a high percentage of bases and an appreciable quantity of illuviated silicate clay; formed on relatively young land surfaces, mostly glacial deposits, from material relatively rich in calcium, under deciduous forests in humid temperate regions.

Great group: The classes in this category contain soils that have the same kind of horizons in the same sequence and have similar moisture and temperature regimes.

Great soil group: Any one of several broad groups of soils with fundamental characteristics in common. Examples are Chernozems, Gray Brown Podzolic, and Podzol.

Green manure: Plant material incorporated with the soil while green, or soon after maturity, for improving the soil.

Ground-Water Laterite soil: A great soil group of the intrazonal order and hydromorphic suborder; consisting of soils characterized by hard-pans or concretional horizons rich in iron and aluminum (and sometimes manganese) that have formed immediately above the water table.

Ground-Water Podzol soil: A great soil group of the intrazonal order and hydromorphic suborder, consisting of soils with an organic mat on the surface aver a very thin layer of acid humus material underlain by a whitish-gray leached layer, which may be as much as 60–90 cm in thickness, and is underlain by a brown, or very dark-brown, cemented hardpan layer; formed under various types of forest vegetation in cool to tropical humid climates under conditions of poor drainage.

Groundwater: Subsurface water in the zone of saturation that is free to move under the influence of gravity.

Gypsic horizon: A mineral soil horizon of secondary calcium sulfate enrichment that is more than 15 cm thick.

Half-Bog soil: A great soil group, of the intrazonal order and hydromorphic suborder, consisting of soil with dark-brown or black peaty material over grayish and rust mottle4 mineral toll; formed under conditions of poor drainage under forest, sedge, or grass vegetation in cool to tropical humid climates.

Hardpan: A hardened soil layer, in the lower A or in the B horizon, caused by cementation of soil particles with organic matter or with material such as sesquioxides, or calcium carbonate. The hardness does not change appreciably with changes in moisture content and pieces of the hard layer do not slake in water. *See also* caliche; claypan; duripan.

Harrowing: A secondary broadcast tillage operation that pulverizes, smooths, and firms the soil in seedbed preparation, controls weeds, or incorporates material spread on the surface.

Heaving: The partial lifting of plants, buildings, roadways. fenceposts, etc., out of the ground, as a result of freezing and thawing of the surface soil during the winter.

Heavy metals: Metals with densities >5.0.

Heavy soil: (Obsolete in scientific use) A soil with a high content of the fine separates, particularly clay, or one with a high drawbar pull, hence difficult to cultivate.

Hemic materials: Intermediate in degree of decomposition of organic materials between the less decomposed fibric and the more decomposed sapric materials.

Heterotroph: An organism capable of deriving energy for life processes only from the decomposition of organic compounds and incapable of using inorganic compounds as sole sources of energy or for organic synthesis. *Contrast with* autotrophic.

Histic epipedon: A thin organic soil horizon that is saturated with water at some period of the year unless artificially drained and that is at or near the surface of a mineral soil.

Histosols: Soils formed from materials high in organic matter. Histosols with essentially no clay must have at least 20% organic matter by weight (about 78% by volume). This minimum organic matter content rises with increasing clay content to 30% (85% by volume) in soils with at least 50% clay.

Holomorphic soil: A. suborder of the intrazonal soil order, consisting of saline and alkali soils formed under imperfect drainage in arid regions and including the great soil group Solonchak or saline soils, Solonetz soils, and Soloth soils. [1949 classification system).

Horizon, soil: A layer of soil, approximately parallel to the soil surface, differing in properities and chacteristics from adjacent layers below or above it. See *also* diagnostic subsurface horizons; diagnostic surface horizons.

Hue: One of the three variables of color. It is caused by light of certain wave-lengths and changes with the wavelength. *See also* chroma; Munsell color system; value, color.

Humic acid: A mixture of variable or indefinite composition of dark organic substances, precipitated upon acidification of a dilute alkali extract from soil.

Humic Gley soil: Soil of the intrazonal order and hydromorphis suborder that includes Wiesenboden and related soils, such as Half-Bog soils, which have a thin muck or peat 02 horizon and an At horizon. Developed in wet meadow and in forested swamps.

Humification: The processes involved in the decomposition of organic matter and leading to the formation of humus.

Humin: The fraction of the soil organic matter that is not dissolved upon extraction of the soil with dilute alkali.

Humus: That more or less stable fraction of the soil organic matter remaining after the major portion of added plant and animal residues have decomposed. Usually it is dark in color.

Hydraulic conductivity: An expression of the readiness with which a liquid such as water flows through a solid such as soil in response to a given potential gradient.

Hydrologic cycle: The circuit of water movement from the atmosphere to the earth and return to the atmosphere through various stages or processes, as precipitation, interception, runoff, infiltration, percolation, storage, evaporation, and transpiration. Hydromorphic soils: A suborder of intrazonal soils, all formed under condition of poor drainage in marshes, swamps, seepage areas, or flats.

Hydroxyapatite: A member of the apatite group of minerals rich in hydroxyl groups. A nearly insoluble calcium phosphate.

Hygroscopic coefficient: The amount of moisture in a dry soil when it is in equilibrium with some standard relative humidity near a saturated atmosphere (about 98%), expressed in terms of percentage on the basis of oven-dry soil.

Igneous rock: Rock formed from the cooling and solidification of magma, and which has not been changed appreciably since its formation.

Illuvial horizon: A soil layer or horizon in which material carried from an overlying layer has been precipitated from solution or deposited from suspension. The layer of accumulation.

Immature soil: A soil with indistinct or only slightly developed horizons because of the relatively short time it has been subjected to the various soil-forming processes. A soil that has not reached equilibrium with its environment.

Immobilization: The conversion of an element from the inorganic to the organic form in microbial tissues or in plant tissues, thus rendering the element not readily available to other organisms or to plants.

Impervious: Resistant to penetration by fluids or impervious by roots.

Inceptisols: Soils that are usually moist with pedogenic horizons of alteration of parent materials but not of illuviation. Generally, the direction of soil development is not yet evident from the marks left by various soil-forming processes or the marks are too weak to classify in another order.

Indurated (Soil): Soil material cemented into a hard mass that will not soften on wetting.

Infiltration rate: A soil characteristic determining or describing the *maximum rate* at which water can enter the soil under specified conditions, including the presence of an excess of water.

Infiltration: The downward entry of water into the soil.

Inoculation: The process of introducing pure *or* mixed cultures of microorganisms into natural or artificial culture media.

Intergrade: A soil that possesses moderately well-developed distinguishing characteristics of two or more genetically related great soil groups.

Interlayer: (mineralogy) Materials between layers within a given crystal, including cations, hydrated cations, organic molecules, and hydroxide groups or sheets.

Intrazonal soils: Soils with more or less well-developed soil characteristics that reflect the dominating influence of some local factor of relief, parent material, or age over the normal effect of climate and vegetation.

Ions: Atoms, groups of atoms, or compounds that are electrically charged as a result of the loss of electrons (cations) or the gain of electrons (anions).

Iron-pan: An indurated soil horizon in which iron oxide is the principal cementing agent.

Irrigation efficiency: The ratio of the water actually consumed by crops on an irrigated area to the amount of water diverted from the source onto the area.

Irrigation methods: Methods by which water is artificially applied to an area. The methods and the manner of applying the water are as follows.

Isomorphous substitution: The replacement of one atom by another of similar size in a crystal lattice without disrupting or changing the crystal structure of the mineral.

Kame: A conical hill or ridge of sand or gravel deposited in contact with glacial ice.

Kaolinite: An aluminosilicate mineral of the 1:1 crystal lattice group; that is, consisting of one silicon tetrahedral sheet alternating with one aluminum octahedral sheet.

Labile: Descriptive of a substance in soil that readily undergoes transformation or is readily available to plants.

Lacustrine deposit: Material deposited in lake water and later exposed either by lowering of the water level or by the elevation of the land.

Land: A broad term embodying the total natural environment of the areas of the earth not covered by water. In addition to sail, its attributes include other physical conditions such as mineral deposits and water supply; location in relation to centers of commerce, populations, and other land; the size of the individual tracts or holdings; and existing plant cover, works of improvement, and the like.

Land capability classification: A grouping of kinds of soil into special units, subclasses, and classes according to their capability for intensive use and the treatments required for sustained use.

Land classification: The arrangement of land units into various categories based upon the properties of the land or its suitability for some particular purpose.

Land-use planning: The development of plans for the uses of land that, over long periods, will best serve the general welfare, together with the formulation of ways and means for achieving such uses.

Lap dressing: An application of fertilizer to a soil after crop stand has been established. A sequence of related soils that differ, one from the other, primarily because of topography as a soil-formation factor.

Lap soil: 1. The layer of soil moved in cultivation. 2. Presumably fertile soil material used to top-dress roadbanks, gardens, lawns.

Laterite: An iron-rich subsoil layer found in some highly weathered humid tropical soils that, when exposed and allowed to dry, becomes very hard and will not soften when rewetted. When erosion removes the overlying layers the laterite is exposed and a virtual pavement results.

Latosol: A suborder of zonal soils including soils formed under forested, tropical, humid conditions and characterized by low silica sesquioxide ratios of the clay fractions, low base exchange capacity, low activity of the clay, low content of most primary minerals, low content of soluble constituents, a high degree of aggregate stability, and usually a red color.

Lattice structure: The orderly arrangement of atoms in a crystalline material.

Layer (clay mineralogy): A combination in silicate clays of (tetrahedral and octahedral) sheets in a 1:1, 2:1, or 2:2 combination.

Leaching: The removal of materials in solution from the soil by percolating waters. See also eluviation.

Legume: A pod-bearing member of the Leguminosae family, one of the most important and widely distributed plant families.

Includes many valuable food and forage species, such as peas, beans, peanuts, clovers, alfalfas. sweet clovers, lespedezas, vetches, and kudzu. Nearly all legumes are nitrogen-fixing plants.

Liebig's law: The growth and reproduction of an organism are determined by the nutrient substance (oxygen, carbon dioxide, calcium, etc.) that is available in minimum quantity, the *limiting factor*.

Light soil (Obsolete in scientific use): A coarse textured soil; a soil with a low drawbar pull and hence easy to cultivate. *See also* coarse texture; soil texture.

Lime (Agricultural): In strict chemical terms, calcium oxide. In practical terms, it is a material containing the carbonates, oxids and/or hydroxides or calcium and/or magnesium used to neutralize, soil acidity.

Limestone: A sedimentary rock composed primarily of calcite ($CaCO_3$). If dolomite ($CaCO_3MgCO_3$) is present in appreciable quantities, it is called a dolomitic limestone.

Lime requirement: The mass of agricultural lime-stone, or the equivalent of other specified liming material required to raise the pH of the soil to a desired value under field conditions.

Liquid fertilizers: Fluid fertilizers that contain essential elements in liquid forms either as soluble nutrients or as liquid suspensions or both.

Liquid limit (LL): The water content corresponding to the arbitrary limit between the liquid and plastic states of consistency of a soil.

Lithosols: A great soil group of azonal soils characterized by an incomplete solum or no clearly expressed soil morphology and consisting of freshly and imperfectly weathered rock or rock fragments.

Loam: The textural class name for soil having a moderate amount of 'sand, silt, and clay. Loam soils contain 7-27% clay, 28-50% silt, and <52% sand.

Loamy: Intermediate in texture and properties between fine-textured and coarse-textured soils. Includes all textural classes with the words 'loam" or "loamy" as a part of the class name, such as clay loam or loamy sand.

Loess: Material transported and deposited by wind and consisting of predominantly silt-sized particles.

Low Humic Gley soils: An intrazonal group of somewhat poorly to poorly drained soils with very thin surface horizons moderately high in organic matter over gray and brown mineral horizons, which are developed under wet conditions.

Luxury consumption: The intake by a plant of an essential nutrient in amounts exceeding what it needs. Thus if potassium is abundant in the soil, alfalfa may take in more that it requires.

Lysimeter: A device for measuring percolation and leaching losses from a column of soil under controlled conditions.

Macronutrient: A chemical element necessary in large amounts (usually 50 ppm in the plant) for the growth of plants. Includes C, H, 0, N, P. K, Ca, Mg, and S. ("Macro' refers to quantity and not to the essentiality of the element).

Marl: Soft and unconsolidated calcium carbonate, usually mixed with varying amounts of clay or other impurities.

Marsh: Periodically wet or continually flooded areas with the surface not deeply submerged. covered dominantly with sedges, cattails, rushes, or other hydrophytic plants. Subclasses include freshwater and salt-water marshes.

Matric potential: That portion of the total soil water potential due to the attractive forces between water and soil solids as represented through adsorption and capillarity. It will always be negative.

Mature soil: A soil with well-developed soil horizons produced by the natural processes of soil formation and essentially in equilibrium with its present environment.

Maximum water-holding capacity: The average moisture content of a disturbed sample of soil, 1 cm high, which is at equilibrium with a water table at its lower surface.

Medium texture: Intermediate between fine-textured and coarse-textured (soils). It includes the following textural classes: very fine sandy loam, loam, silt loam, and silt.

Mellow soil: A very soft, very friable, porous soil without any tendency toward hardness or harshness.

Metamorphic rock: A rock that has been greatly altered from its previous condition through the combined action of heat and pressure. For example, marble is a metamorphic rock produced from limestone; gneiss is produced from granite and slate from shale.

Micas: Primary aluminosilicate minerals in which two silica tetrahedral sheets alternate with one alumina octahedral sheet. They separate readily into thin sheets or flakes.

Microfauna: That part of the animal population which consists of individuals too small to be clearly distinguished without the use of a microscope. Includes protozoans and nematodes.

Microflora: That part of the plant population which consists of individuals too, small to be clearly distinguished without the use of a microscope. Includes actinomycetes, algae, bacteria, and fungi.

Micronutrient: A chemical element necessary in only extremely small amounts (<50 ppm in the plant) for the growth of plants. Examples are B, Cl, Cu, Fe, Mn, and Zn. ("Micro" refers to the amount used rather than to its essentiality).

Microrelief: Small-scale local differences in topography, including mounds, swales, or pits that are only a meter or so in diameter and with elevation differences of up to 2 m.

Mineral soil: A soil consisting predominantly of, and having its properties determined predominantly by, mineral matter. Usually contains <20% organic matter, but may contain an organic surface layer up to 30 cm thick.

Mineralization: The conversion of an element from an organic form to an inorganic state as a result of microbial decomposition.

Minimum tillage: The minimum soil manipulation necessary for crop produc-tion or meeting tillage requirements under the existing soil and climatic conditions.

Mixed fertilizers: Two or more fertilizers materials mixed together. May be as dry powders, granules, pellets, bulk blends, or liquids. The quantity of certain plant nutrient elements needed, in addition to the amount supplied by the soil, to increase plant growth to a designated optimum.

Moderately coarse texture: Consisting predominantly of coarse particles. In soil textural classification, it includes all the sandy loams except the very fine sandy loam. *See also* coarse texture.

Moderately fine texture: Consisting predominantly of intermediate-sized (soil) particles or with relatively small amounts of fine or coarse particles. In soil textural classification, it includes clay loam, sandy loam, sandy clay loam, and silty clay loam.

Moisture equivalent: The weight percentage of water retained by a previously saturated sample of soil 1 cm in thickness after it has been subjected to a centrifugal force of 1000 gravity for 30 min.

Moisture tension (pressure): The equivalant negative pressure in the soil water. It is equal to the equivant pressure that must be applied to the soil water to bring it to hydraulic equilibrium, through a porous permeable wall or membrane, with a pool of water of the same composition.

Mole drain: Unlined drain formed by pulling a bullet-shaped cylinder through the soil.

Mollic epipedon: A surface horizon of mineral soil that is dark colored and relatively thick, contains at least 0.58% organic carbon, is not massive and hard or very hard when dry, has a base saturation of more than 50% when measured at pH 7, has less than 250 ppm of P_2O_5, soluble in 1% citric acid, and is dominantly saturated with bivalent cations.

Mollisols: Soils with nearly black, organic- rich surface horizons and high supply of bases. They have mollic epipedons and base saturation greater than 50% in any cambic or argillic horizon. They lack the characteristics of Vertisols and must not have oxic or spodic horizons.

Montmorillonite: An aluminosilicate clay mineral in the smectite group with a 2:1 expanding crystal lattice, with two silicons tetrahedral sheets enclosing an aluminum octahedral sheet. Considerable expansion may be caused by water moving between silica sheets of continuous layers.

Mor: Raw humus; a type of forest layer of unincorporated organic material, usually matted or compacted or both; distinct from the mineral soil, unless the latter has blackened by washing in organic matter.

Moraine: An accumulation of drift, with an initial topographic expression of its own within a glaciated region chiefly by the direct action of glacial ice. Examples are ground, lateral, recessional, and terminal moraines.

Morphology, soil: The constitution of the soil including the texture, structure, consistence, color, and other physical, chemical, and biological properties of the various soil horizons that makes up the soil profile.

Mottling: Spots or blotches of different color shades of color interspersed with the dominant color

Muck soil: 1. A soil containing 20-50% organic matter. 2. An organic soil in which the organic matter is well decomposed.

Muck: Highly decomposed organic material in which the original plant parts are not recognizable. Contains more mineral matter and is usually darker in color than peat.

Mulch tillage: Tillage or preparation of the soil in such a way that plant residues or other materials are left to cover the surface; also called *mulch forming, trash farming, stubble mulch tillage, plowless forming.*

Mulch: Any material such as straw, sawdust, leaves, plastic film, and loose soil that is spread upon the surface of the soil to protect the soil and plant roots from the effects of raindrops, crusting, freezing, evaporation etc.

Mull: A humus-rich layer of forested soils consisting of mixed organic and mineral matter. A mull blends into the upper mineral-layers F without an abrupt change in soil characteristics.

Munsell color system: A color designation system that specifies the relative degree of the three simple variables of color: hue, value, and chroma. For example: 10YR 6/4 is a color (of soil) with a hue = 10YR, value = 6, and chrome = 4. These notations can be translated into several different systems of color names as desired.

Mycorrhiza: The association, usually symbiotic, of fungi with the roots of seed plants. *See also* ectotrophic mycorrhiza; endotrophic mycorrhiza; vesicular-arbuscular mycthrhiza.

Natric horizon: A mineral soil horizon that satisfies the requirements of an argillic horizon, but that also has prismatic, columnar, or blocky structure and a subhorizon having more than 15% saturation with exchangeable sodium.

Natural erosion: Wearing away of the earth's surface by water, ice or other natural agents under natural environmental conditions of climate, vegetation, and so on, undisturbed by man. Synonymous with *geological erosion.*

Necrosis: Death associated with discoloration and dehydration of all or parts of plant organs, such as leaves.

Nematodes: Very small worms abundant in many soils and important because some of them attack and destroy plant roots.

Neutral soil: A soil in which the surface layer, at least to normal plow depth, is neither acid nor alkaline in reaction.

Nitrification: The biochemical oxidation of ammonium to nitrate, predominately by autotrophic bacteria.

Nitrogen assimilation: The incorporation of nitrogen into organic cell substances by living organisms.

Nitrogen cycle: The sequence of chemical and biological changes undergone by nitrogen as it moves from the atmosphere into water, soil, and living organisms, and upon death of these organisms (plants and animals) is recycled through part or all of the entire process.

Nitrogen fixation: The biological conversion of elemental nitrogen (N_2) to organic combinations or to forms readily utilizable in biological processes.

Noncalcic brown soils: The zonal group of soils with slightly acid, light pinkish or light reddish brown A horizons over light reddish brown or dull red B horizons developed under mixed grass and forest vegetation in a subhumid, wet dry climate.

No-tillage system: A procedure whereby crop is planted directly into a seedbed not tilled since harvest of the previous crop; also zero tillage.

Nucleic acids: Complex compounds found in plant and animal cells; may be combined with proteins as nucleoproteins.

O horizon: Organic horizon of mineral soils.

O: structureless-no observable aggregation. 1. Weakly durable peds. 2. Moderately durable peds. 3. Strong, durable peds.

Ochric epipedon: A surface horizon of mineral soil that is too light in color, too high in chroma, too low in organic carbon, or too thin to be a plaggen, mollic, umbric. anthropic, or histic epipedon, or that is both hard and massive when dry.

Order: The category at the highest level of generalization in the soil classification system. The properties selected to distinguish the orders are reflections of the degree of horizon development and the kinds of horizons present. The ten orders are as follows.

Organic fertilizer: By-product from the processing of animal or vegetable substances that contain sufficient plant nutrients to be of value as fertilizers.

Organic soil materials: (As used in *Soil Taxonomy* in the United States): 1. Saturated with water for prolonged periods unless artificially drained and having 18% or more organic carbon (by weight) if the mineral fraction is more than 60% clay, more than 12% organic carbon if the mineral fraction has no clay, or between 12 and 18% carbon if the clay content of the mineral fraction is between 0 and 60%. 2. Never saturated with water for more than a few days and having more than 20% organic carbon. Histosols develop on these organic soil materials. There are three kinds of organic materials.

Organic soil: A soil that contains at least 20% organic matter (by weight) if theclay content is low and at least 30% if the clay contents as high as 60%.

Ortstein: An indurated layer in the B horizon of Podzols in which the cementing material consists of illuviated sesquioxides (mostly iron) and organic matter.

Osmotic potential: That portion of the total soil water potential due to the presence of solutes in soil water. It will generally be negative.

Osmotic: A type of pressure exerted in living bodies as a result of unequal concentration of salts on both sides of a cell wall or membrane. Water will move from the area having the least salt concentration through the membrane into the area having the highest salt concentration and, therefore, exerts additional pressure on this side of the membrane.

Oven-dry soil: Soil that has been dried at 105°C until it reaches constant weight.

Oxic horizon: A mineral soil horizon that is at least 30 cm thick and characterized by the virtual absence of weatherable primary minerals or 2:1 lattice clays and the presence of 1:1 lattice clays and highly insoluble minerals such as quartz sand, hydrated oxides of iron and aluminum, low cation exchange capacity, and small amounts of exchangeable bases.

Oxidation ditch: An artificial open channel for partial digestion of liquid organic wastes in which the wastes are circulated and aerated by a mechanical device.

Oxisols: Soils with residual accumulations of inactive clays, free oxides, kaolin, and quartz. They are mostly in tropical climates.

Pans: Horizons or layers, in soils, that are strongly compacted, indurated, or very high in clay content. *See also* caliche; claypan; duripan; fragipan; hardpan.

Parent material: The unconsolidated and more or less chemically weathered mineral or organic matter from which the solum of soils is developed by pedogenic processes.

Particle density: The mass per unit volume of the soil particles. In technical work, usually expressed as grams per cubic centimeter (g/cm^3) or metric tons per cubic meter (Mg/m^3).

Particle size analysis: Determination of the various amounts of the different separates in a soil sample, usually by sedimentation, sieving, micrometry, or combinations of these methods.

Particle size distribution: The amounts of the various soil separates in a soil sample, usually impressed as weight percentages.

Particle size: The effective diameter of a particle measured by sedimentation, sieving, or micro-metric methods.

Peat soil: An organic soil containing more than 40% organic matter. Used in the United States to refer to the stage of decomposition of the organic matter, "peat" referring to the slightly decomposed. or undecomposed deposits and "muck" to the highly decomposed materials. *See also* muck; muck soil; peat.

Peat: Unconsolidated soil material consisting largely at undocomposed, or only slightly decomposed organic matter accumulated under conditions at excessive moisture. *See also* organic soil materials; peat soil.

Ped: A unit of soil structure such as an aggregate, crumb, prism, block, or granule, formed by natural processes (in contrast with a clod, which is formed artificially).

Pedon: The smallest volume that pan be called "a soil," it has three dimensions. It extends downward to the depth of plant roots or to the lower limit of the genetic soil horizons. Its lateral cross section is roughly hexagonal and ranges from 1 to 10 m^2 in size depending on the variability in the horizons.

Peneplain: A once high, rugged area that has been reduced by erosion to a low, gently rolling surface resembling a plain.

Penetrability: The ease with which a probe can be pushed into the soil. May be expressed in units of distance, speed, force, or work depending on the type of penetrometer used.

Percolation, soil water: The downward movement of water through soil. Especially, the downward flow of water in saturated or nearly saturated soil at hydraulic gradients of the order of 1.0 or less.

Permafrost: 1. Permanently frozen material underlying the solum. 2. A perennially frozen soil horizon.

Permanent charge: The net negative (or positive) charge of clay particles inherent in the crystal lattice of the particle; not affected by changes in pH or by ion-exchange reactions.

Permeability, soil: The ease with which gases, liquids, or plant roots penetrate or pass through a bulk mass of soil or a layer of soil. pF: (Obsolete) The logarithm of the soil moisture tension expressed in centimeters height of a column of water.

Petrocalcic horizon: A continuous, indurated calcic horizon that is cemented by calcium carbonate and, in some places, with magnesium carbonate. It cannot be penetrated with a spade or auger when dry, dry fragments do not slake in water, and it is impenetrable to roots.

Petrogypsic horizon: A continuous, strongly cemented, massive gypsic horizon that is cemented by calcium sulfate. It can be chipped with a spade when dry. Dry fragments do not slake in water and it is impenetrable to roots.

pH, soil: The negative logarithm of the hydrogen ion activity of a soil. The degree of acidity (or alkalinity) of a soil as determined by means of a glass, quinhydrone, or other suitable electrode or indicator at a specified moisture content or soil-water ratio, and expressed in terms of the pH scale.

pH-dependent charge: That portion of the total charge of the soil particles that is affected by and varies with, changes in pH.

Phase, soil: A subdivision of a soil series or other unit of classification having characteristics that affect the use and management of the soil but that do not vary sufficiently to differentiate it as a separate series. A variation in a property or characteristic such as degree of slope, degree of erosion, and content of stones.

Photomap: A mosaic map made from aerial photographs with physical and cultural features shown as on a planimetric map.

Physical properties (of soils): Those characteristics, processes, or reactions of a soil that are caused by physical forces and that can be described by, or expressed in, physical terms or equations. Examples of physical properties are bulk density, water-holding capacity, hydraulic conductivity, porosity, pore-size distribution, and so on.

Placic horizon: A black to dark reddish mineral soil horizon that is usually thin but that may range from 1 mm to 25 mm in thickness. The placic horizon is commonly cemented with iron and is slowly permeable or impenetrabl to water and roots.

Plaggen epipedon: A man-made surface horizon more than 50 cm thick that is formed by long-continued manuring and mixing. Umbric epipedon. A surface layer of mineral soil that has the same requirements as the mollic epipedon with respect to color, thickness, organic carbon content, consistence, structure, and P_2O_5, content, but that has a base saturation of less than 50% when measured at pH 7.

Planosol: A great soil group of the intrazonal order and hydromorphic suborder consisting of soils with eluviated surface horizons underlain by B horizons more strongly eluviated, cemented, or compacted than associated normal soil.

Plastic limit (PL): The water content corresponding to an arbitrary limit between the plastic and semisolid states of consistency of a soil.

Plastic soil: A Boil capable of being molded or deformed continuously and permanently, by relatively moderate pressure, into various shapes. *See also* consistence.

Platy: Consisting of soil aggregates that are developed predominantly along the horizontal axes; laminated; flaky.

Plinthite (brick): A highly weathered mixture of sesquioxides or iron and aluminum with quartz and other diluents that occurs as red, mottles and that changes irreversibly to hardpan upon alternate wetting and drying.

Plow layer: The soil ordinarily moved when land is plowed; equivalent to surface soil.

Plow pan: A subsurface soil layer having a higher bulk density and lower total porosity than layers above or below it. as a result of pressure applied by normal plowing and other tillage operations.

Plowing: A primary broad-base tillage operation that is performed to shatter soil uniformly with partial to complete inversion.

Plow-plant: *See* tillage, conservation.

Plow-planting: The plowing and planting of a land in a single trip over the field by drawn in both plowing and planting tools with the same power sources.

Podzol: A great soil group of the zonal order consisting of soils formed in cool-temperate to temperate, humid climates, under coniferous or mixed coniferous and deciduous forest, and characterized particularly by a highly leached, whitish gray A2 horizon.

Podzolization: A process of soil formation resulting in the genesis of Podzols and podzolic soils.

Polypedon: Two or more contiguous pedons, all of which are within the defined limits of a single soil series. In early stages of the development of the classification scheme this was called a soil individual.

Pore size distribution: The volume of the various sizes of pores in a soil. Expressed as percentages of the bulk volume (soil plus pore space).

Porosity, soil: The volume percentage of the total soil bulk not occupied by solid particles.

Prairie soils: A zonal great soil group consisting of soils formed under temperate to cool-temperate, humid regions under tall grass vegetation.

Primary mineral: A mineral that has not been altered chemically since deposition and crystallization from molten lava.

Prismatic soil structure: A soil structure type with prism-like aggregates that have a vertical axis much longer than the horizontal axes.

Productivity, soil: The capacity of a soil for producing a specified plant or sequence of plants under a specified system of management. Productivity emphasizes the capacity of soil to produce crops and should be expressed in terms of yields.

Profile, soil: A vertical section of the soil through all its horizons and extending into the parent material.

Protein: Any of a group of nitrogen-containing organic compounds formed by the polymerization of a large number of amino acid molecules and that, upon hydrolysis, yield these amino acids. They are essential parts of living matter and are one of the essential food substances of animals.

Puddled soil: Dense, massive soil artificially compacted when wet and having no regular structure. The condition commonly results from the tillage of a clayey soil when it is wet.

Reddish brown laterite soils (of U.S): A zonal group of coil with dark reddish brown granular surface soils rod friable clay B horizons, and red or reticulately mottled lateritic parent material, developed under humid tropical climate with wet-dry seasons and tropical forest vegetation.

Reddish Brown soils: A zonal group of soils with a light brown surface horizon of a slightly reddish cast which grades into dull reddish brown or red material heavier than the surface soil, then into a horizon of whitish or pinkish lime accumulation developed under shrub and shortgrass vegetation of warm-temperate to tropical regions of semiarid climate.

Red-Yellow Podzolic soils: A zonal great soil group consisting of soils formed under warm temperate to tropical, humid climates, under deciduous or coniferous forest vegetation and usually under conditions of good drainage.

Regolith: The unconsolidated mantle of weathered rock and soil material on the earth's surface; loose earth materials above solid rock. (Approximately equivalent to the term "soil" as used by many engineers).

Regosol: Any soil of the azonal order without definite genetic horizons and developing from or on deep. Unconsolidated, soft mineral deposits such as sands, loess, or glacial drift.

Rendzina: A great soil group of the intrazonal order and calcimorphic suborder consisting of soils with brown or black friable surface horizons underlain by light gray to pale yellow calcareous material; developed from soft, highly calcareous parent material under grass vegetation or mixed grasses and forest in humid and semiarid climates.

Residual material: Unconsolidated and partly weathered mineral materials accumulated by disintegration of consolidated rock in place.

Rhizobia: Bacteria capable of living symbiotically with higher plants, usually in nodules on the roots of legumes, from which they receive their energy, and capable of converting atmospheric nitrogen to combined organic forms hence, the term symbiotic nitrogen-fixing bacteria. (Derived from the generic name *Rhizobium*).

Rhizosphere: That portion of the soil in the immediate vicinity of plant roots in which the abundance and composition of the microbial population are influenced by the presence of roots.

Rid Dosed soil: A zonal great soil group consisting of soils formed under warm temperate to hot, dry regions under desert-type vegetation, mostly shrubs.

Rill erosion: An erosion process in which numerous small channels of only several centimeters in depth are formed; occurs mainly on recently cultivated soils.

Rill: A small, intermittent water course with steep sides; usually only a few inches deep and hence no obstacle to tillage operations.

Riprap: Broken rock, cobbles, or boulders placed on earth surfaces, such as the face of a dam or the bank of a stream, for protection against action of water (waves); also applied to brush or pole mattresses, or brush and stone, or other similar materials used for soil erosion control.

Rock: The material that forms the essential part of the earth's solid crust, including loose incoherent masses such a sand and gravel, as well as solid masses of granite and limestone.

Runoff: The portion of the precipitation on an area that is discharged from the area through stream channels. That which is lost without entering the soil is called surface runoff and that which enters the soil before reaching the stream is called groundwater runoff or seepage flow from groundwater. (In soil science "runoff" usually refers to the water lost by surface flow; in geology and hydraulics "runoff" usually. includes both surface and subsurface flow).

Saline soil: A nonsodic soil containing sufficient soluble salts to impair its productivity.

Saline-sodic soil: A soil containing sufficient exchangeable sodium to interfere with the 'growth of most crop plants and containing appreciable quantities of soluble salts. The exchangeable-sodium percentage is >15, the conductivity of the saturation extract >2 ds/m (at 25°C), and the pH is usually 8.5 or less in the saturated soil.

Salinization: The process of accumulation of salts in soil.

Salk horizon: A mineral soil horizon of enrichment with secondary salts more soluble in cold water than gypsum. A salic horizon is 15 cm or more in thickness.

Saltation: Particle movement in water or wind where particles skip or bounce along the stream bed or soil surface.

Sand: A soil particle between 0.05 and 2.0 mm in diameter a soil textural class.

Sapric materials: The most highly decomposed of the organic materials, having the highest bulk density, least amount of plant fiber, and lowest water content at saturation.

Savanna (savannah): A grassland with scattered trees, either as individuals or clumps. Often a transitional type between true grassland and forest.

Second bottom: The first terrace above the normal flood plain of a stream.

Secondary mineral: A mineral resulting from the decomposition of a primary mineral or from the reprecipitation of the products of decomposition of a primary mineral.

Sedimentary rock: A rock formed from materials deposited from suspension or precipitated from solution and usually being more or less consolidated. The principal sedimentary rocks are sandstones, shales, limestones, and conglomerates.

Self-mulching soil: A soil in which the surface layer becomes so well aggregated that it does not crust and seal under the impact of rain but instead serves as a surface mulch upon drying.

Separate, soil: One of the individual-sized groups of mineral oil particles-sand, silt or clay,

Series: The soil series is a subdivision of a family and consists of soils that are similar in all major profile characteristics.

Sewage sludge: Settled sewage solids combined with varying amounts of water and dissolved materials, removed from sewage by screening, sedimentation, chemical precipitation, or bacterial digestion.

Shear: Force, as of a tillage implement, acting at right angles to the direction of movement.

Sheet: (minerology) A flat array of more than one atomic thickness and composed of one or more levels of linked coordination polyhedra. A sheet is thicker than a plane and thinner than a layer. Example: tetrahedral sheet, octahedral sheet.

Sheet erosion: The removal of a fairly uniform layer of soil from the land surface by runoff water.

Shelterbelt: A wind harrier of living trees and shrubs established and maintained for protection of farm fields. Syn. windbreak.

Shrinkage limit (SL): The maximum water content at which a reduction in water content will not cause a decrease in the volume of the soil mass. This defines the arbitrary limit between the solid and semisolid states.

Sierozem: A zonal great soil group consisting of soils with pale-grayish A horizons grading into calcareous material at a depth of 30 cm or less, and formed in temperate to cool arid climates under a vegetation of desert plants, short grass, and scattered brush.

Silica/alumina ratio: The molecules of silicon dioxide (S_iO_2) per molecule of aluminum oxide (Al_2O_3) in clay minerals or in soils.

Silica/sesquioxide ratio: The molecules of silicon dioxide (S_iO_2) pet molecule of aluminum oxide(Al_2O_3)plus ferric oxide (Fe_2O_3) in clay minerals or in soils.

Silt: 1. A soil separate consisting of particles between 0.05 and 0.002mm in equivalent diameter. 2. A soil textural class.

Silting: The deposition of water-borne sediments in stream channels, lakes, reservoirs, or on flood plains, usually resulting from a decrease in the velocity of the water.

Site Index: A quantitative evaluation of the productivity of a soil for forest growth under the existing or specified environment.

Slag: The product of smelting, containing mostly silicates; the substances not sought to be produced as matte or metal and having a lower specific gravity.

Slick spots: Small areas in a field that are slick when wet, due to a high content of alkali or exchangeable sodium.

Slope: The degree of deviation of a surface from horizontal, measured in a numerical ratio, percent, or degrees.

Smectite: A group of silicate clays having a 2:1 type lattice structure with sufficient isomophorous substitution in either or both the tetrahedral and octahedral sheets to give a high inter-layer negative charge and high cation exchange capacity and to permit significant interlayer expansion and consequent shrinking and swelling of the clay. Montmorillonite, beidellite, and saponite are in the smectite group.

Sod planting: A method of planting in sod with little or no tillage.

Sodic soil: A soil that contains sufficient sodium to interfere with the growth of most crop plants, and in which the exchangeable sodium percentage is 15 or greater.

Sodium adsorption ratio (SAR).

$$SAR = \frac{[Na^{+}]}{\sqrt{\frac{[Ca^{+2}] + [Mg^{+2}]}{2}}}$$

where the cation concentrations are in millimoles per liter (mmole/l).

Soil air: The soil atmosphere; the gaseous phase or the soil, being that volume not occupied by solid of liquid.

Soil alkalinity: The degree or intensity of alkalinity of a soft, expressed by a value >7.0 on the pH scale.

Soil amendment: Any material, such as lime, gypsum, sawdust, or synthetic conditioner, that is worked into the soil to make it more amenable to plant growth.

Soil association: A group of defined and named taxonomic soil units occurring together in an individual and characteristic pattern over a geographic region, comparable to plant associations in many ways.

Soil classification: *(Soil Taxonomy]* The systematic arrangement of soils into groups or categories on the basis of their characteristics. The categories of the system used in the U.S. since 1968 are described briefly.

Soil complex: A mapping unit used in detailed soil surveys where two or more defined taxonomic units are so intimately. Intermixed geographically that it is undesirable or impractical, because of the scale being used, to separate them. A more intimate mixing of smaller areas of individual taxonomic units than that described under soil association.

Soil conditioner: Any material added to a soil for the purpose of improving its physical condition.

Soil conservation: A combination of all management and land-use methods that safeguard the soil against depletion or deterioration caused by nature and/or humans.

Soil correlation: The process of defining, mapping, naming, and classifying the kinds of soils in a specific soil survey area, the purpose being to ensure that soils are adequately defined, accurately mapped, and uniformly named.

Soil genesis: The mode of origin of the soil, with special reference to the processes or soil-forming factors responsible for the development of the solum, or true soil, from the unconsolidated parent material.

Soil geography: A subspecialization of physical geography concerned with the areal distributions of soil types.

Soil management: The sum total of all tillage operations, cropping practices, fertilizer, lime, and other treatments conducted on or applied to a soil for the production of plants.

Soil map: A map showing the distribution of soil types or other soil mapping units in relation to the prominent physical and cultural features of the earth's surface.

Soil mechanics and engineering: A subspecialization of soil science concerned with the effect of forces on the soil and the application of engineering principles to problems involving the soil.

Soil microbiology: A subspecialization of soil science concerned with soil inhabiting microorganisms and with their relation to agriculture, Including both plant and animal growth.

Soil moisture potential: A vertical section of a soil profile removed from the soil and mounted for display or study.

Soil morphology: The physical constitution, particularly the structural properties, of a soil profile as exhibited by the kinds, thickness, and arrangement of the horizons in the profile, and by the texture, structure, consistency, and porosity of each horizon.

Soil organic matter: The organic fraction of the soil that includes plant and animal residues at various stages of decomposition, cells and tissues of soil organisms, and substances synthesized by the soil population. Commonly determined as the amount of organic material contained in a soil sample passed through a 2-mm sieve.

Soil profile: A vertical section of the soil from the surface through all its horizons, including C horizons.

Soil reaction: Measure the activity of the Hydrogen ions of a solution.

Soil salinity: The amount of soluble salts in a soil, expressed in terms of percentage, parts per million (ppm), or other convenient ratios. Soil solution: The aqueous liquid phase of the soil and its solutes consisting of ions dissociated from the surfaces of the soil particles and of other soluble materials.

Soil structure classes: A grouping of soil structural units or peds on the basis of size from the very fine to very coarse.

Soil structure grades: A grouping or classification of soil structure on the basis of inter and intraaggregate adhesion, cohesion, or stability within the profile. Four grades of structure, designated from 0 to 3 are recognized.

Soil structure types: A classification of soil stricture based on the shape of the aggregates or peds and their arrangement in the profile, including platy, prismatic, columnar, blocky, subangular blocky, granulated, and crumb.

Soil structure: The combination or arrangement of primary soil particles into secondary particles, units, or peds. These secondary units may be, but usually are not, arranged in the profile in such a manner as to give a distinctive characteristic pattern. The secondary units are characterized end classified on the basis of size, shape, and degree of distinctness into classes, types, and grades respectively.

Soil survey: The systematic examination, description, classification, and mapping of soils in an area. Soil surveys are classified according to the kind and intensity of field examination.

Soil temperature classes: (Soil *Taxonomy)* Classes are based on mean annual soil temperature and on differences between summer and winter temperatures at a depth of 50 cm.

1. Soils with 5°C and greater difference between summer and winter temperatures are classed on the basis of mean annual temperatures frigid: <8°C mean annual temperature.

   mesic: 8-15°C mean annual temperature.

   thermic 15-22°C mean annual temperature.

   hyperthennic: >22°C mean annual temperature.

2. Soils with <5°C difference between summer and winter temperatures are classed on the basis of mean annual temperatures.

   isofrigid: <8°C mean annual temperature.

   isomesic: 8-15°C mean annual temperature.

   Isothermic: 15-22°C mean annual temperature.

   Isohyperthermic: >22°C mean annual temperature.

Soil textural class: A grouping of soil textural units based on the relative proportions of the various soil separates (sand, silt, and clay). These textural classes, listed from the coarsest to the finest in texture, are sand, loamy sand, sandy loam, loam, silt loam, silt, sandy clay loam, clay loam, silty clay loam, sandy clay, silty clay, and clay. There are several subclasses of the sand, loamy sand, and sandy loam classes based on the dominant particle size of the sand fraction (e.g., loamy fine sand, coarse sandy loam).

Soil texture: The relative proportions of the various soil separates in a soil.

Soil type: The lowest unit in the natural system of soil classification; a sub div-sion of a soil series and consisting of or describing soils that are alike in all characteristics including the texture of the A horizon.

Soil water potential: (total) A measure of the difj ference in the free energy state of soil water: and that of pure water. Technically it is defined as "that amount of work that must be done per unit quantity of pure water In order to transport reversibly and isothermically an infinitesimal quantity of water from a pool of pure water, at a specified elevation and at atmospheric pressure to the soil water (at the point under consideration)," This *total* potential consists of the following potentials.

Soil: 1. A dynamic natural body composed of mineral and organic materials and living forms in which plants grow. 2. The collection of natural bodies occupying parts of the earth's surface that support plants and that have properties due to the integrated effect of climate and living matter acting upon parent material, as conditioned by relief, over periods of time.

Sol brun Acide: A zonal group of soils developed under forest vegetation with thin At horizon, a paler A2 horizon that is poorly differentiated from the B2 horizon, a B2 horizon with uniform color from top to bottom, weak subangular blocky structure, and lacking evidence of silicate clay accumulation. The sola are strongly to very strongly acid.

Solian soil material: soil material accumulated through wind action. The most extensive areas is the United States are silty deposits (loess), but large areas of sandy deposits also occur.

Solodized soil: A soil that has been subjected to the processes responsible for the development of a Soloth and having at least some of the characteristics of a Soloth.

Solonchak soils: An intrazonal group of soils with high concentrations of soluble salts inrelation to those in other soils, usually light-colored, without characteristic structural form, developed under salt-loving plants, and occurring mostly in a subhumid or semiarid climate.

Solonetz soils: An intrazonalgroup of soils having surface horizons of varying degrees of friability underlain by dark hard soil, ordinarily with columnar structure (prismatic structure with rounded tops). This hard layer is usually highly alkaline. Such soils are developed under grass or shrub vegetation, mostly in subhumid or semiarid climates. Soluble-sodium percentage (SSP). The proportion of sodium ions in solution in relation to the total cation concentration, defined as :

$$SSP = \frac{\text{Soluble sodium conc. [cmol/L]}}{\text{Total cation conc. [cmol/L]}} \times 100$$

Solum (plural: sola): The upper and most weathered part of the soil profile; the A and B horizons.

Splash erosion: The spattering of small soil particles caused by the impact of raindrops on very wet soils. The loosened and separated particles may or may not be subsequently removed by surface runoff.

Spodic horizon: A mineral soil horizon characterize by the illuvial accumulation of amorphous materials composed of aluminum and organic carbon with or without iron.

Spodosols: Soils with subsurface illuvial accumulations of organic matter and compounds of aluminum and usually iron. These soils are formed in acid, mainly coarse-textured materials in humid and mostly cool or temperate climates.

Sprinkler: The water is sprayed over the soil surface through nozzles from a pressure system.

Stratified: Arranged in or composed of strata or layers.

Strip cropping: The practice of growing crops that require different types of tillage, such as row and sod, in alternate strips along contours or across the prevailing direction of wind.

Stubble mulch: The stubble of crops or crop residues left essentially in place on the land as a surface cover before and during the preparation of the seedbed and at least partly during the growing of a succeeding crop.

Sub irrigation: The water is applied in open ditches or tile lines until the water table is raised sufficiently to wet the soil.

Subgroup: The great groups are subdivided into central concept subgroups that show the central properties of the great group, inter-grade subgroups that show properties of more than one great group, and other subgroups for soils with a typical properties that are not characteristic of any great group. family. Families are defined largely on the basis of physical and mineralogic properties of importance to plant growth.

Suborder: This category narrows the ranges in soil moisture and temperature regimes, kinds of horizons, and composition, according to which of these is most important.

Subsoil: A tillage implement with one or more cultivator type feet to which are attached strong knifelike units used to shatter or loosen hard layers, usually in the subsoil, to below normal plow depth.

Subsoil: That part of the soil below the plow layer.

Subsoiling: Breaking of compact subsoils, without inverting them, with a special knife like instrument (chisel), which is pulled through the soil at depths, usually of 30-60 cm and at spacing's usually of 1-2 m.

Subsurface tillage: Tillage with a special sweep-like plow or blade that is drawn beneath the surface, cutting plant roots and loosening the soil without inverting it or without incorporating residues of the surface cover.

Summer fallow: The tillage of uncropped land during the summer in order to control weeds and store moisture in the soil for the growth of a later crop. surface runoff.

Surface soil: The uppermost part of the soil, ordinarily moved in tillage. or its equivalent in uncultivated soils. Ranges in depth from 7 to 25 cm. Frequently designated as the "plow layer," the "Ap layer," or the "Ap horizon."

Symbiosis: The living together in intimate association of two dissimilar organisms, the cohabitation being mutually beneficial.

Talus: Fragments of rock and other soil material accumulated by gravity at the foot of cliffs or steep slopes.

Taxonomy, soil: The science of classification of soils; laws and principles governing the classifying the soil.

Tensiometer: A device for measuring the negative pressure (or tension) of water in soil *in situ:* a porous, permeable ceramic cup connected through a tube to a manometer or vacuum gauge.

Terrace: 1. A level, usually narrow, plain bordering a river, lake, or the sea. Rivers sometimes are bordered by terraces at different levels. 2. A raised, more or less level or horizontal strip of earth usually constructed on or nearly on a contour and designed to make the land suitable for tillage and to prevent accelerated erosion by diverting water from undesirable channels of concentration.

Tertiary waste treatment: Waste-water treatment beyond the secondary or biological stage that includes removal of nutrients such as phosphorus and nitrogen and of a high percentage of suspended solids.

Thermal analysis (differential thermal analysis): A method of analyzing a soil sample for constituent based on a differential rate of heating of the unknown end standard samples when a uniform source of heat is applied.

Thermophilic organisms: Organisms that gin readily at temperatures above 45°C.

Tile, drain: Pipe made of burned clay, concrete or ceramic material, in short lengths usually laid with open joints to collect and carry excess water from the soil.

Till: 1. Unstratified glacial drift deposited rectly by the ice and consisting of clay, sand gravel, and boulders intermingled in any portion. 2. To plow and prepare for seeding to seed or cultivate the soil.

Tillage: The mechanical manipulation of soil is any purpose; but in agriculture it is usually restricted to the modifying of soil conditions ii crop production.

Tillage, conservation: Any tillage sequence h reduces loss of soil or water relative to conventional tillage, including the following system.

Tillage, conventional: The combined primary and secondary tlllage operations normally performed in preparing a seedbed for a given crop in a given geographic area.

Tillage primary: Tillage that contributes to the major soil manipulation.

Tillage rotary: An operation using a power driven rotary tillage tool to loosen and mix soil.

Tilth: The physical condition of soil as related to its ease of tlllage, fitness as a seedbed, and its impedance to seedling emergence and root penetration.

Ultisols: Soils that are low in bases, and have subsurface horizons of illuvial clay accumulations. They are usually moist, but during the warm season of the year some are dry part of the time.

Unsaturated flow: The movement of water in a soil that is not filled to capacity with water.

Value, color: The relative lightness or intensity of color and approximately a function of the square root of the total amount of light. One of the three variables of color.

Varnish, desert: A glossy sheen or coating on stones and gravel in arid regions.

Vertisols: Clayey soils with high shrink swell potential that have wide, deep cracks when dry. Most of these soils have distinct wet and dry periods throughout the year.

Vesicular-arbbscular mycorrhiza: A common endomycorrhizal association produced by phycomycetous fungi of the genus *Endogone*. Host range includes many agricultural and horticultural crops.

Waterlogged: Saturated with water.

Virgin soil: A soil that has not been significantly disturbed from Its natural environment.

Water table, perched: The surface of a local zone of saturation held above the main body of groundwater by an impermeable layer of stratum, usually clay, and separated from the math body of groundwater by an unsaturated zone.

Water table: The upper surface of ground water or that level below which the soil is saturated with water.

Water use efficiency: Dry matter or harvested portion of crop produced per unit of water consumed.

Water-stable aggregate: A soil aggregate stable to the action of water such as falling drops, or agitation as in wet-sieving analysis.

Weathering: All physical and chemical changes produced in rocks, at or near the earth's surface, by atmospheric agents.

Wheel track planting: A practice of planting in which the seed is planted in tracks formed by wheels rolling immediately ahead of the planter.

Wild-flooding: The water is released at high points in the field and distribution is uncontrolled.

Wilting point (permanent wilting point): The moisture content of soil, on an oven-dry basis, at which plants wilt and fail to recover their turgidity when placed in a dark humid atmosphere.

Xerophytes: Plants that grow in or on extremely dry soils or soil materials.

Zonal soil: A soil characteristic of a large area or zone.

□□□